# Undergraduate Probability

## A Brief Introduction

# Frank Blume

Cover credit: © Yeyendesign | Dreamstime.com

ISBN-10: 1519161654
ISBN-13: 978-1519161659

*To*
*Ferdi and Babette*

# Acknowledgments

I am grateful to John Brown University for supporting my work on this book by granting my request for a sabbatical leave in the fall of 2015. Special thanks go to my colleagues Calvin Piston and Gregory Varner who kindly agreed to teach some of my courses during my absence.

# *Preface*

*Undergraduate Probability* is meant to be a textbook for a one-semester introductory college-level course. In my own teaching experience, the amount of material covered in this book is somewhat larger than the average student can be expected to process in a single three-hour course. But this excess in volume and topical variety is likely beneficial because it gives instructors using this text greater flexibility in course design and subject selection. Sections that can be considered optional are

- Section 2.10 on ideal gases and statistical mechanics,
- Section 2.11 on climate change and correlation,
- Section 3.1 on the weak law of large numbers in the binomial case,
- Section 3.3 on the strong law of large numbers,
- Section 4.2 on randomness and consciousness, and
- Section 4.3 on regression lines and scatter plots.

Furthermore, Sections 4.5 and 4.6 on estimation by means of chi-squared and $T$ densities may have to be left out due to time constraints.

Conceptually and stylistically, this text aims to favor depth over breadth and thus provides proofs or cogent justifications—if at all possible—for all the propositions, theorems, and applications that are included in it. Moreover, the central role played by the normal distribution and the *Central Limit Theorem* is strongly emphasized, and relevant applications are carefully explained—especially in Sections 2.7, 2.10, and 2.11.

As far as prerequisites are concerned, it is important to point out that a solid background in Calculus I and II is a must and that knowledge of multivariable calculus is highly desirable.

That said, there only remains for me to express my hope that students who diligently study this book will be rewarded in their labors by gaining genuine, dependable insight, and will come to appreciate a challenging subject that vitally enriches and informs a vast variety of modern fields of knowledge

FRANK BLUME, NOVEMBER 2015.

# Contents

# Chapter 1

# Probability Basics

## 1.1 Sets and Functions

Before we can begin our study of the theory of probability, we need to briefly review some basic definitions and facts concerning sets and functions. Roughly speaking, a *set* is an arbitrary collection of objects of our thought or perception. Sets may consist of apples and oranges as well as of cars and chairs or just about anything else that we can somehow conceive of. In mathematics, of course, apples and oranges are not that commonly encountered, but numbers definitely are. Thus we say, for example, that $\{1, 2, 3\}$ is the set containing the numbers $1, 2$, and $3$. The curly brackets are standard notation indicating that we are dealing with a set. We also refer to the numbers $1, 2$, and $3$ as the *elements* of the set $\{1, 2, 3\}$. In general, if an object $a$ is an element of a set $A$, we express this relation by writing $a \in A$ (e.g., $2 \in \{1, 2, 3\}$, but *not* $0 \in \{1, 2, 3\}$). Furthermore, a set $A$ is said to be a *subset* of a set $B$, written as $A \subset B$, if every element of $A$ is also an element of $B$ (e.g., $\{1, 2\} \subset \{1, 2, 3\}$ but *not* $\{0, 2\} \subset \{1, 2, 3\}$).

The following list of definitions establishes the most important operations with sets and provides simple illustrations in each case:

a) The *union* $A \cup B$ of two sets $A$ and $B$ is the set of all elements that are contained in $A$ *or* $B$.
   Examples: $\{1, 5\} \cup \{2, 3, 4\} = \{1, 2, 3, 4, 5\}$ and $\{1, 5\} \cup \{1, 2\} = \{1, 2, 5\}$.

b) The *intersection* $A \cap B$ consists of all elements that are contained in both $A$ *and* $B$.
   Examples: $\{1, 3, 5\} \cap \{1, 2, 3\} = \{1, 3\}$, $\{1, 2\} \cap \{1, 2, 3\} = \{1, 2\}$, and $\{1, 2\} \cap \{3, 4\} = \emptyset$, where we denote by '$\emptyset$' the empty set that contains no elements at all.

c) The *difference* $A \setminus B$ is the set of all elements that are contained in $A$

1

but not in $B$.

Examples: $\{1, 2, 3\} \smallsetminus \{1, 4\} = \{2, 3\}$ and $\{1, 2, 3\} \smallsetminus \{4, 5\} = \{1, 2, 3\}$.

**d)** The *Cartesian product* $A \times B$ is the set of all ordered pairs $(a, b)$ of elements $a \in A$ and $b \in B$.

Examples: $\{1, 2, 3\} \times \{1, 4\} = \{(1, 1), (2, 1), (3, 1), (1, 4), (2, 4), (3, 4)\}$ and $\{1, 2, 3\} \times \emptyset = \emptyset$.

One of the sets that we encounter most frequently is the set $\mathbb{R}$ of all *real numbers*. Intuitively, we think of a real number as a point on a 'continuous' coordinate axis, and some examples of real numbers are $1$, $-1/3$, $0.356$, $\sqrt{2}$, and $\pi$. A commonly occurring type of subset of $\mathbb{R}$ is the *interval*: for $a, b \in \mathbb{R}$ with $a < b$, the interval from $a$ to $b$ is the set of all real numbers between $a$ and $b$. What we always need to be a little careful about is the inclusion or exclusion of the *boundary points* $a$ and $b$. Inclusion is indicated by a bracket and exclusion by a parenthesis. Thus, there are four possible types of intervals:

$$[a, b], \; [a, b), \; (a, b], \; \text{and} \; (a, b).$$

The first in the list is said to be a *closed* interval, the next two are called *half-open* and the last is referred to as *open*. For instance, $[-3, 5)$ is the half-open interval that contains all real numbers greater than or equal to $-3$ and strictly less than $5$. By convention, we also allow $\pm\infty$ as 'values' for $a$ and/or $b$. For instance, the interval $[0, \infty)$ is the set of all real numbers greater than or equal to $0$, $(-\infty, 1)$ is the set of all real numbers strictly less than $1$, and $(-\infty, \infty)$ is equal to all of $\mathbb{R}$.

Apart from $\mathbb{R}$ and subsets of $\mathbb{R}$, there also are the higher-dimensional Cartesian products of $\mathbb{R}$ with itself:

$$\mathbb{R}^2 := \mathbb{R} \times \mathbb{R} = \{(x, y) \mid x \in \mathbb{R} \wedge y \in \mathbb{R}\},$$
$$\mathbb{R}^3 := \mathbb{R} \times \mathbb{R} \times \mathbb{R} = \{(x, y, z) \mid x \in \mathbb{R} \wedge y \in \mathbb{R} \wedge z \in \mathbb{R}\},$$

and in general,

$$\mathbb{R}^n := \{(x_1, \ldots, x_n) \mid x_1 \in \mathbb{R} \wedge \cdots \wedge x_n \in \mathbb{R}\},$$

for any positive integer $n$ (where '$\wedge$' means 'and'). Finally, the most prominent set of *integers* that we will be working with is the set

$$\mathbb{N} = \{1, 2, 3, \ldots\}$$

of positive integers, and occasionally we also encounter the set

$$\mathbb{Z} = \{\ldots, -2, -1, 0, 1, 2, \ldots\}$$

of all integers—positive, negative, or zero.

**1.1.1 Example.** The Cartesian product $[0,1] \times [2,4)$ is the set of all pairs $(x,y) \in \mathbb{R}^2$ for which $0 \le x \le 1$ and $2 \le y < 4$, that is,

$$[0,1] \times [2,4) = \{(x,y) \in \mathbb{R}^2 \mid 0 \le x \le 1 \wedge 2 \le y < 4\}.$$

**1.1.2 Example.** The set $S = \{(x,y) \in \mathbb{R}^2 \mid x^2 + y^2 \le 4\}$ is the disc of radius 2 centered at the origin $(0,0) \in \mathbb{R}^2$ (see Figure 1.1). This is so because for every point $(x,y) \in \mathbb{R}^2$ the quantity $x^2 + y^2$ is the square of the distance of $(x,y)$ from the origin $(0,0)$.

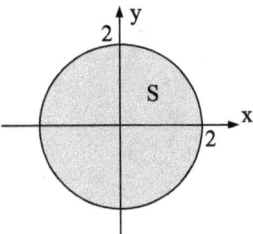

Figure 1.1: the disc $S$.

**1.1.3 Example.** The set $S = \{(x,y) \in \mathbb{R}^2 \mid |x-y| \le 1 \wedge |x+y| \le 1\}$ is the square with vertices at $(1,0)$, $(0,1)$, $(-1,0)$, and $(0,-1)$ (as shown in Figure 1.2) because the inequalities $|x-y| \le 1$ and $|x+y| \le 1$ are equivalent to

$$y \le x + 1 \wedge y \ge x - 1$$

and

$$y \le 1 - x \wedge y \ge -1 - x,$$

respectively.

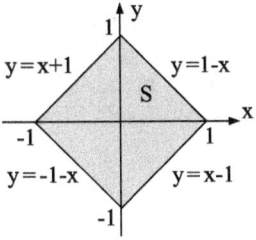

Figure 1.2: the square $S$.

Having thus reviewed the notion of a set, we wish to point out that there are numerous instances in mathematics in general and in the theory of probability in particular when we consider elements of one set in dependence on or

correspondence to elements of another set. Certain types of such correspon-
dences are referred to as *functions*, and the concept of a function is extremely
important. Indeed, it is more fundamental to mathematics than even the concept of a number. To give an example, we may consider the equation $y = x^2$.
An equation of this form can be interpreted as a *rule of correspondence* or
*rule of assignment* in the sense that it specifies for a given input value $x$ the
output value $y$ to be $x^2$. However, for a complete definition of a function we
need to specify not only a functional relation between input and output values
(possibly represented by an equation), but also the sets that these input and
output values are taken from.

**1.1.4 Definition.** Let $A$ and $B$ be nonempty sets. A *function f* from $A$ to
$B$ is a rule of correspondence that assigns to each element in $A$ *exactly one*
element in $B$. The standard notation that symbolizes the dependence of an
*output $f(x)$* in $B$ on an *input $x$* in $A$ is

$$f : A \to B$$
$$x \mapsto f(x).$$

Furthermore, $A$ is said to be the *domain* of $f$, and $R(f) := \{f(x) \mid x \in A\}$ is
the *range* of $f$.

**1.1.5 Example.** The equation $y = x^2$ can be interpreted as a rule of assignment in the sense that every value $x$ is assigned the value $x^2$. So we may
write

$$x \mapsto x^2.$$

However, in order to define a function, a rule of assignment alone is not
sufficient—we also need to say exactly which values are permitted as input
for $x$. In other words, we need to specify the domain $A$. There are many possible definitions for $A$, and which of these we pick may depend on the larger
context, but for simplicity we will in this example choose $A$ to be the set $\mathbb{R}$ of
all real numbers. Then the output values $x^2$ are real numbers as well, and we
may therefore set $B := \mathbb{R}$ (the colon in front of the equal sign indicates that $B$
is *defined* to be $\mathbb{R}$). Having thus specified $A$ and $B$, we introduce the following
function:

$$f : \mathbb{R} \to \mathbb{R}$$
$$x \mapsto x^2. \tag{1.1}$$

The range $R(f)$ is in this case *not* equal to $B = \mathbb{R}$, because the square of any
number $x \in \mathbb{R}$ is always greater than or equal to zero. In other words, the only
possible output values of $f$ are the positive real numbers including zero, that
is,

$$R(f) = [0, \infty).$$

For clarity we wish to point out that it would not have been wrong to replace $B = \mathbb{R}$ in the definition of $f$ with $[0, \infty)$ so that (1.1) would read

$$f : \mathbb{R} \to [0, \infty)$$
$$x \mapsto x^2.$$

However, in working with functions it is usually sufficient to be given only the general type rather than the exact range of output values. So the less precise notation in (1.1) is not only acceptable but typically preferable. Moreover, for simplicity we will frequently replace the element-assignment notation '$x \mapsto x^2$' with the more familiar looking '$f(x) := x^2$' (again the colon in front of the equal sign indicates that $f(x)$ is *defined* to be $x^2$). So instead of (1.1) we will also write, "$f : \mathbb{R} \to \mathbb{R}$ is a function defined by the equation $f(x) := x^2$."

A special type of function that we often come across is the *sequence* which is a function defined on $\mathbb{N}$. For instance, the function $a : \mathbb{N} \to \mathbb{R}$, $a(n) := 1/n$ is the infinite sequence of the real numbers

$$a(1) = 1, \ a(2) = 1/2, \ a(3) = 1/3, \ a(4) = 1/4, \ldots \ ad \ infinitum.$$

Traditionally, however, the function notation $a(n)$ is here replaced by the sequence notation $a_n$, and instead of $a : \mathbb{N} \to \mathbb{R}$ we usually write $(a_n)_{n \in \mathbb{N}}$.

Commonly occurring sequences in the context of probability are not only sequences of real numbers, but also sequences of functions and sets. For example, if $f_n(x) := x^2/n$ for all $x \in \mathbb{R}$ and all $n \in \mathbb{N}$ and $I_n := [0, 1/n]$ for all $n \in \mathbb{N}$, then $(f_n)_{n \in \mathbb{N}}$ is a sequence of functions, and $(I_n)_{n \in \mathbb{N}}$ is sequence of sets, or more precisely, of intervals. Moreover, if $(A_n)_{n \in \mathbb{N}}$ is sequence of sets, then

$$\bigcup_{n=1}^{\infty} A_n := \{s \mid s \in A_n \text{ for some } n \in \mathbb{N}\}$$

and

$$\bigcap_{n=1}^{\infty} A_n = \{s \mid s \in A_n \text{ for all } n \in \mathbb{N}\}$$

are the union and the intersection of the sets in this sequence, respectively.

**1.1.6 Example.** Setting $A_n := [0, 1-1/n]$ and $B_n := [0, 1+1/n]$ for all $n \in \mathbb{N}$, it follows that

$$\bigcup_{n=1}^{\infty} A_n = [0, 0] \cup [0, 1/2] \cup [0, 2/3] \cup [0, 3/4] \cup \cdots = [0, 1)$$

and

$$\bigcap_{n=1}^{\infty} B_n = [0, 2] \cap [0, 3/2] \cap [0, 4/3] \cap [0, 5/4] \cap \cdots = [0, 1].$$

## Exercises

**1.1.7.** Which of the three sets $\{1, 2, 3, 4\}$, $\{1, 2, 5\}$, and $\{1, 2\}$ are subsets of $\{1, 2, 3, 4\}$?

**1.1.8.** Identify the elements in each of the following sets: $\{1, 2\} \cup \{2, 5, 6\}$, $\{1, 2, 3, 4\} \cap \{2, 4, 7\}$, and $\{1, 2, 3, 4\} \setminus \{2, 4, 7\}$.

**1.1.9.** Which of the numbers $-2$, $-1$, $1/2$, $1$, and $3/2$ are contained in the interval $[-1, 1)$?

**1.1.10.** Write each of the following sets as a single interval: $[1, 2) \cup [2, 3]$, $[1, 2] \setminus \{2\}$, $[1, 2] \setminus \{1, 2\}$, $[0, 3] \setminus [1, 3]$, $\mathbb{R} \setminus (-\infty, 0)$, $[1, 2] \cap (3/2, 3)$, $[1, 2] \cap [2, 3]$, and $(a, b) \cup (b, a)$ for $a, b \in \mathbb{R}$ with $a \neq b$.

**1.1.11.** Determine the range of the function $f : [-1, 2] \to \mathbb{R}$, $x \mapsto 2x - 1$.

**1.1.12.** Let a function $f$ be defined by the equation $f(x) := \dfrac{x - 3}{(2 - x)\sqrt{x^2 + 3}}$.

    a) What is the largest possible domain of $f$ (as a subset of $\mathbb{R}$)?

    b) Find the values $f(3)$ and $f(-3)$.

    c) Find the value $f(f(1))$.

**1.1.13.** What is the largest possible domain of the function $f(x) := \sqrt{x^2 - 4}$ (as a subset of $\mathbb{R}$)?

**1.1.14.** Is it possible to use the equation $y^2 = x^2$ as a rule of correspondence that defines $y$ as a function of $x$ with domain $\mathbb{R}$? Explain your answer. *Hint:* according to Defintion 1.1.4, a function from $A$ to $B$ assigns to each element in $A$ *exactly* one element in $B$.

**1.1.15.** For each of the definitions given below find $\bigcup_{n=1}^{\infty} A_n$.

    a) $A_n := \{1, \ldots, n\}$,

    b) $A_n := [1/n - 1, 1/n]$,

    c) $A_n := [1/n, \arctan(n)]$,

    d) $A_n := \{(x, y) \in \mathbb{R}^2 \mid x^2 + y^2 \leq 1 - 1/n\}$.

**1.1.16.** For each of the definitions given below find $\bigcap_{n=1}^{\infty} A_n$.

    a) $A_n := \{1, \ldots, n\}$,

    b) $A_n := [0, 1/n]$,

    c) $A_n := [-1 - 1/n, 1 + 1/n]$,

    d) $A_n := \{(x, y) \in \mathbb{R}^2 \mid x^2 + y^2 \leq 1/n\}$

**1.1.17.** For each of the definitions given below, decide whether $A \subset B$, $B \subset A$, or $A = B$.

   **a)** $A := \mathbb{N}$, $B := \bigcup_{k=1}^{\infty}\{k\}$,

   **b)** $A := \mathbb{Z}$, $B := \mathbb{N}$,

   **c)** $A := \{1\}$, $B := \emptyset$,

   **d)** $A := (-\infty, \infty) \times \mathbb{R}$, $B := \mathbb{R}^2$.

**1.1.18.** For the function shown in Figure 1.3 answer the following questions:

   **a)** What are the domain and range of $f$?

   **b)** What is the value of $f$ at $x = 2$?

   **c)** For which value(s) of $x$ is $f(x)$ equal to 3?

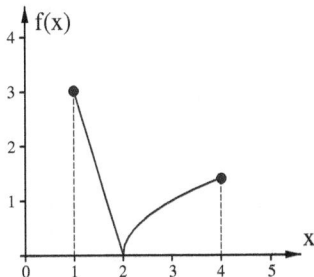

Figure 1.3: graph of $f$.

**1.1.19.** Sketch each of the following subsets of $\mathbb{R}^2$:

   **a)** $\{(x, y) \in \mathbb{R}^2 \mid y \geq x^2\}$,

   **b)** $\{(x, y) \in [0, 1] \times [0, 1] \mid xy \leq 1/3\}$,

   **c)** $\{(x, y) \in [0, 1] \times [0, 1] \mid x^2 y \leq 1/9\}$,

   **d)** $\{(x, y) \in \mathbb{R}^2 \mid |y| + |x| \leq 1\}$.

## 1.2 Sample Spaces

Roughly speaking, the theory of probability aims to bring to light the mathematical laws that underly physical random processes such as the tossing of a coin or the rolling of a die. Consequently, the first and most elementary object of study to which this theory pertains is the set of outcomes that a given random experiment can produce. Hence we introduce the following definition:

**1.2.1 Definition.** The *sample space* of a random experiment, usually denoted by $S$, is the set of all possible outcomes that the experiment can generate.

**1.2.2 Example.** If a coin is tossed once, and the outcome is recorded as 'head' $(H)$ or 'tail' $(T)$, then $S = \{H, T\}$.

**1.2.3 Example.** If a coin is tossed twice, and the outcome is recorded as an ordered pair of letters, then $S = \{HH, HT, TH, TT\}$.

The next example shows that the sample space is not uniquely determined by the physical process that is used to conduct a certain random experiment: we can toss a coin twice and consider the pairs in the set $\{HH, HT, TH, TT\}$ to be the corresponding outcomes or we can, say, count the number of heads in each pair and construe the sample space $S$ to consist of these numbers.

**1.2.4 Example.** If a coin is tossed twice, and the outcome is recorded by counting the number of heads, then $S = \{0, 1, 2\}$.

**1.2.5 Example.** If a coin is tossed $n$ times, and the outcome of each toss is recorded as 'head' $(H)$ or 'tail' $(T)$, then $S$ consists of all the ordered sequences of length $n$ in the letters $H$ and $T$. If $n = 2$, then $S = \{HH, HT, TH, TT\}$ (see Example 1.2.3), and if $n = 3$, then

$$S = \{HHH, HHT, HTH, HTT, THH, THT, TTH, TTT\}.$$

So in passing from $n = 2$ to $n = 3$, the number of elements in $S$ doubles from $4 = 2^2$ to $8 = 2^3$. The reason for this doubling is that each pair in the set $\{HH, HT, TH, TT\}$ can be followed by either an $H$ or a $T$ in case that $n = 3$. Consequently, since every triple in the sample space above, for $n = 3$, can again be followed by an $H$ or a $T$ as $n$ is increased to 4, we may infer that the number of outcomes for $n = 4$ doubles from $8 = 2^3$ to $16 = 2^4$. In general, therefore, the number of outcomes in $S$ is equal to $2^n$ if the number of tosses is $n$.

**1.2.6 Example.** If a coin is tossed $n$ times, and the number of heads is recorded, then $S = \{0, 1, \ldots, n\}$.

**1.2.7 Example.** If a die is rolled once and the number it shows is recorded, then $S = \{1, 2, 3, 4, 5, 6\}$.

**1.2.8 Example.** If a die is rolled once and an $A$ is written down if the number on the die is a 1 or a 2 and a $B$ is written down otherwise, then $S = \{A, B\}$.

**1.2.9 Example.** If a die is rolled twice and the sum of the numbers it shows is recorded, then $S = \{2, 3, \ldots, 12\}$

## Exercises

**1.2.10.** A probability experiment is performed by rolling two dice—one blue, one black—and its outcome is determined by dividing the number on the blue die by the number on the black die. In other words, the outcome of the experiment is a single number obtained by dividing two given numbers. Find the sample space of this experiment.

**1.2.11.** A die is rolled three times, and the three numbers that the die shows are recorded. How many outcomes are there in the sample space of this experiment?

**1.2.12.** A die is rolled three times, and the sum of the numbers that it shows is recorded. What is the sample space of this experiment?

**1.2.13.** A number $k$ is chosen at random from the set $\{1, \ldots, 10\}$ and then a number $i$ is chosen at random from the set $\{k, \ldots, 10\}$. Find the sample space of this experiment.

**1.2.14.** A number $k$ is chosen at random from the set $\{1, \ldots, 10\}$, a number $i$ is chosen at random from the set $\{1, \ldots, k\}$, and the sum $k + i$ is recorded. Find the sample space of this experiment.

**1.2.15.** Find the sample space in the preceding exercise if the sum $k + i$ is replaced by the product $ki$.

# 1.3 Events and Probabilities

When we speak of a probability in a well defined mathematical sense, we *always* speak of a probability of an *event*. Probabilities are probabilities of events and nothing else—there are no exceptions.

**1.3.1 Definition.** An *event* is a subset of a sample space.

**1.3.2 Example.** The empty set—denoted by the symbol '$\emptyset$'—is a subset of every set and is therefore, in particular, an event in every sample space. Since the empty set contains no elements—or outcomes—it is said to be the *impossible event*.

**1.3.3 Example.** If a coin is tossed once, and the outcome is recorded as 'head' ($H$) or 'tail' ($T$), then $S = \{H, T\}$ and the events contained in $S$ are $\emptyset$, $\{H\}$, $\{T\}$, and $\{H, T\}$.

**1.3.4 Example.** If a coin is tossed twice, and the outcome is recorded by counting the number of heads, then $S = \{0, 1, 2\}$ and the events contained in $S$ are $\emptyset$, $\{0\}$, $\{1\}$, $\{2\}$, $\{0, 1\}$, $\{0, 2\}$, $\{1, 2\}$, and $\{0, 1, 2\}$.

**1.3.5 Theorem.** *If the number of elements in a sample space $S$ is $n$ (i.e., $\#S = n$), then the number of events in $S$ is $2^n$.*

*Proof.* Assume that $S = \{s_1, \ldots, s_n\}$ and let $E \subset S$ be a given event. Then $E$ is uniquely characterized by the function

$$f_E : S \to \{0, 1\}$$

$$s_k \mapsto \begin{cases} 1 & \text{if } s_k \in E, \\ 0 & \text{if } s_k \notin E, \end{cases}$$

in the following way: the event $E$ consists of exactly those elements in $S$ for which the value of $f_E$ is equal to 1. Given this unique identification of events $E$ with functions $f_E$, it follows that the number of events $E$ is equal to the number of functions $f_E$. Furthermore, since every function $f_E$ assigns to every element in $S$ either a one or a zero, we may identify $f_E$ with the zero-one sequence of its output values: $(f_E(s_1), \ldots, f_E(s_n))$. But as in Example 1.2.5, where we found the number of sequences of length $n$ in the letters $H$ and $T$ to be $2^n$, so we find here as well that the number of zero-one sequences of length $n$ is equal to $2^n$. □

**1.3.6 Definition.** If $S$ is a sample space of a given random experiment and $E \subset S$ is an event, then the probability for an outcome in $S$ to be an element of $E$ is said to be the *probability of $E$* and is denoted by $P(E)$. Furthermore, a probability always is a number between 0 and 1, that is, $P(E) \in [0,1]$ for all events $E \subset S$, and the total probability of all the outcomes combined is always equal to 1, that is, $P(S) = 1$.

*Remark.* As indicated in the preceding definition and as we pointed out before, *probabilities are always understood to be probabilities of events.* Consequently, the probability of a single outcome is to be interpreted as the probability of the event consisting only of that single outcome. In other words, if $s \in S$, then the 'probability of $s$' is, strictly speaking, not $P(s)$ but rather $P(\{s\})$. However, this notational distinction is fairly tedious to maintain consistently, and for convenience we will therefore often use the simpler form $P(s)$. All the same though it is important to understand that probabilities universally are probabilities of events. There is no other meaning that the word 'probability' will ever be endowed with in this text.

**1.3.7 Example.** If a perfectly unbiased coin is tossed once and the outcome is recorded as 'head' ($H$) or 'tail' ($T$), then $P(H) = P(\{H\}) = 1/2$, $P(\emptyset) = 0$, and $P(\{H, T\}) = 1$.

**1.3.8 Example.** If a perfectly unbiased die is rolled once and the number it shows is recorded (i.e., $S = \{1, 2, 3, 4, 5, 6\}$), then $P(1) = P(\{1\}) = 1/6$ and $P(\{1, 2\}) = 1/3$.

**1.3.9 Theorem.** *If $S$ is a finite sample space (i.e., if the number of elements in $S$ is finite), and if all outcomes in $S$ occur with equal probability, then for any event $E \subset S$, the probability of $E$ is equal to the number of elements in $E$ divided by the number of elements in $S$, that is,*

$$P(E) = \frac{\#E}{\#S}.$$

*Proof.* Since the total combined probability of all the outcomes in the sample space is equal to one and since all outcomes are equally likely, it follows that the probability of any single outcome is equal to $1/\#S$ and that therefore the probability of an event $E \subset S$ must be equal to $\#E/\#S$ as claimed. □

*Remark.* The 'proof' just given is not really a proof but rather an intuitive justification because a rigorous proof would have to appeal to the general axiomatic properties that the function $P$—as a measure of probability—must be assumed or shown to possess (see Section 1.4).

**1.3.10 Example.** If an unbiased die is rolled once, then $S = \{1, 2, 3, 4, 5, 6\}$ and

$$P(\{1, 3, 6\}) = \frac{\#\{1, 3, 6\}}{\#\{1, 2, 3, 4, 5, 6\}} = \frac{3}{6} = \frac{1}{2}.$$

**1.3.11 Example.** If an unbiased coin is tossed $n$ times, and the outcome is recorded by writing down a sequence of length $n$ in the letters $H$ and $T$, then for all $k \in \{0, 1, \ldots, n\}$ the probability of the event $E_{n,k}$ that consists of all sequences with $k$ heads and $n - k$ tails is

$$P(E_{n,k}) = \frac{\#E_{n,k}}{\#S} = \frac{\#E_{n,k}}{2^n}.$$

This is so because, according to Example 1.2.5, the sample space $S$ of the coin-tossing experiment here in question consists of $2^n$ outcomes and because the assumption that the coin is unbiased implies that all the outcomes in $S$ are equally likely. In order to find a formula for $\#E_{n,k}$, it is helpful to consider an example. So let's assume that $n = 5$ and $k = 3$. Given this choice, we readily find that the event $E_{5,3}$ consists of a total of 10 sequences of length five in the letters $H$ and $T$. Written vertically, these sequences look as follows:

| H | H | H | H | H | H | T | T | T | T |
|---|---|---|---|---|---|---|---|---|---|
| H | H | H | T | T | T | H | H | H | T |
| H | T | T | H | H | T | H | H | T | H |
| T | H | T | H | T | H | H | T | H | H |
| T | T | H | T | H | H | T | H | H | H |

Consequently, the probability of $E_{5,3}$ is

$$P(E_{5,3}) = \frac{10}{2^5} = \frac{10}{32}.$$

In order to generalize this result to the case of $k$ heads and $n - k$ tails, we need to determine in how many different ways we can arrange $k$ letters $H$ and $n - k$ letters $T$ in a row of length $n$. To approach this problem, it is helpful to address the following simpler question first: in how many different ways can we arrange $n$ *distinct* objects in a row of length $n$? Suppose, for example, that we are given three squares labeled 1, 2, and 3. Then, for the first position we can choose either one of the three squares. Having filled the first position, we are left with two choices for the second position, and having filled the first two positions, there is only one choice remaining for the last position. Therefore, the total number of arrangements or *permutations* of three distinct objects is $3 \cdot 2 \cdot 1 = 6$.

| 1 | 1 | 2 | 2 | 3 | 3 |
|---|---|---|---|---|---|
| 2 | 3 | 1 | 3 | 1 | 2 |
| 3 | 2 | 3 | 1 | 2 | 1 |

In general, for an arbitrary positive integer $n$, the number of permutations is $n \cdot (n-1) \cdots 3 \cdot 2 \cdot 1$. By convention, the product $n \cdot (n-1) \cdots 3 \cdot 2 \cdot 1$ is referred to as $n$ *factorial* and is denoted by $n!$. Thus, we have established the following rule:

> The number of permutations of $n$ distinct objects is $n!$.

(Note: for the special case $n = 0$ it is customary to define $0! := 1$.) In order to apply the rule above to the problem of finding the number of arrangements of $k$ letters $H$ and $n-k$ letters $T$, we resort to a little trick: considering again the example of three letters $H$ and two letters $T$ with $n = 5$, we write the letters $H$ and $T$ on five little squares that are labeled with the numbers from 1 to 5.

$$\boxed{H_1}\ \boxed{H_2}\ \boxed{H_3}\ \boxed{T_4}\ \boxed{T_5}$$

Then there are $5! = 120$ different permutations in the indices from 1 to 5, but looking only at the letters $H$ and $T$, some of these permutations are identical. For instance, the permutations

$$\boxed{H_1}\ \boxed{T_4}\ \boxed{H_2}\ \boxed{H_3}\ \boxed{T_5} \tag{1.2}$$

and

$$\boxed{H_3}\ \boxed{T_5}\ \boxed{H_1}\ \boxed{H_2}\ \boxed{T_4} \tag{1.3}$$

display the same sequence in the letters $H$ and $T$, but different permutations in the numbers from 1 to 5. Since there are $3!$ ways to arrange three distinct objects (the squares with an $H$) and $2!$ ways to arrange two distinct objects (the squares with a $T$), we may infer that there are $3! \cdot 2!$ permutations of the numbers from 1 to 5 that display the same sequence $HTHHT$ as in (1.2) and (1.3) above. Given this observation, it is easy to see that the total number of permutations in the letters $H$ and $T$ is

$$\frac{5!}{3! \cdot 2!} = 10.$$

Furthermore and by direct extension, the number of permutations of $k$ letters $H$ and $n - k$ letters $T$ must be $n!/(k!(n - k)!)$. This quotient is referred to as a *binomial coefficient* (see the remark below) and is denoted by $\binom{n}{k}$. Thus we may conclude that the number of permutations of two objects in a row of length $n$ with $k$ times the first object and $n - k$ times the second is

$$\#E_{n,k} = \binom{n}{k} = \frac{n!}{k!(n-k)!}$$

and that the probability of $E_{n,k}$ (in the case of a perfectly unbiased coin) is

$$P(E_{n,k}) = \frac{1}{2^n}\binom{n}{k}.$$

*Remark.* In order to understand why $\binom{n}{k}$ is said to be a 'binomial coefficient', we need to take a look at binomial powers of the form $(a+b)^n$:

$$
\begin{aligned}
(a+b)^0 &= 1 \\
(a+b)^1 &= 1a + 1b \\
(a+b)^2 &= 1a^2 + 2ab + 1b^2 \\
(a+b)^3 &= 1a^3 + 3a^2b + 3ab^2 + 1b^3 \\
(a+b)^4 &= 1a^4 + 4a^3b + 6a^2b^2 + 4ab^3 + 1b^4.
\end{aligned}
$$

Extracting the numerical coefficients from the terms on the right-hand side yields a triangular array known as *Pascal's Triangle*:

$$
\begin{array}{ccccccccc}
 & & & & 1 & & & & \\
 & & & 1 & & 1 & & & \\
 & & 1 & & 2 & & 1 & & \\
 & 1 & & 3 & & 3 & & 1 & \\
1 & & 4 & & 6 & & 4 & & 1 \\
 & \vdots & & & \vdots & & & \vdots &
\end{array}
$$

By inspection we find that the numbers in each row are built by adding the diagonally adjacent numbers in the preceding row: $2 = 1 + 1$, $3 = 1 + 2$, $4 = 1 + 3$, $6 = 3 + 3$, etc. As it turns out, this defining additive property is properly reflected in following general property (see Exercise 1.3.24 below):

$$\binom{n+1}{k} = \binom{n}{k-1} + \binom{n}{k} \tag{1.4}$$

for all positive integers $n$ and all $k \in \{1, \dots, n\}$. Furthermore, as we take a closer look at how the coefficients of $(a+b)^n$ are formed, there is revealed the same combinatorial pattern that we encountered above in Example 1.3.11. For instance, for $n = 3$ we find that

$$
\begin{aligned}
(a+b)^3 &= (a+b)(a+b)(a+b) \\
&= a\cdot a\cdot a + a\cdot a\cdot b + a\cdot b\cdot a + b\cdot a\cdot a \\
&\quad + a\cdot b\cdot b + b\cdot a\cdot b + b\cdot b\cdot a + b\cdot b\cdot b \\
&= a^3 + 3a^2b + 3ab^2 + b^3.
\end{aligned}
$$

The term $a^2 b$, for example, appears three times because three is the number of arrangements of two letters $a$ and one letter $b$ in a row of length three: $aab$, $aba$, and $baa$. Thus it becomes apparent that in general the binomial $(a+b)^n$ is a sum of products of the form $a^k b^{n-k}$ that each appear $\binom{n}{k}$ times. Given this observation, we arrive at the following algebraic formula which is known as the *binomial formula:*

$$(a+b)^n = \sum_{k=0}^{n} \binom{n}{k} a^k b^{n-k}. \tag{1.5}$$

**1.3.12 Theorem.** *If $S$ is a finite sample space, then for any event $E \subset S$ it is the case that*

$$P(E) = \sum_{s \in E} P(s).$$

*Note: here we do not assume all outcomes in $S$ to be equally likely.*

*Proof.* If $S$ consist of finitely many outcomes then so does $E$, and therefore, the probability of $E$ is the finite sum of the probabilities of the outcomes that $E$ comprises. (Note: this 'proof' again is not so much a proof as it is an intuitive justification.)  □

**1.3.13 Example.** If an unbiased coin is tossed $n$ times and an outcome is recorded by writing down the number of heads, then $S = \{0, 1, \ldots, n\}$, and, using the result of Example 1.3.11, it follows that for all $k \in \{0, 1, \ldots, n\}$ the probability of the outcome $k$ is

$$P(k) = P(\{k\}) = \frac{1}{2^n} \binom{n}{k}.$$

Furthermore, according to Theorm 1.3.12, the probability of the event $E = \{0, 1, 2, 3\}$ (for $n \geq 3$) is

$$P(E) = \sum_{k=0}^{3} \frac{1}{2^n} \binom{n}{k}.$$

**1.3.14 Example.** An unbiased die is rolled $n$ times. Whenever the die shows a 1 or a 2 we write down an $A$ and otherwise a $B$. Given this rule, it follows that a single outcome is a sequence of length $n$ in the letters $A$ and $B$. Furthermore, since the probability for writing down an $A$ is $2/6 = 1/3$ and the probability for writing down a $B$ is $4/6 = 2/3$, we may conclude that the probability of any outcome that displays $k$ times an $A$ and $n-k$ times a $B$ is

$$\left(\frac{1}{3}\right)^k \left(\frac{2}{3}\right)^{n-k}.$$

To better understand why this is so, let us consider the sequence $AABBA$ as an example. The probabilty for an $A$ to be written down after the first trial is, as we just said, $1/3$. By implication, if the experiment of rolling a die five times is repeated again and again, then we may expect one third of all the sequences of length 5 in the letters $A$ and $B$ thus produced to display an $A$ in the first position. Moreover, out of this latter set of outcomes we may expect another third to display an $A$ again in the second position. Consequently, the probability for an outcome to display two $A$'s in the first two positions is $(1/3)^2$. Continuing in this manner, we next infer that we may expect two thirds of all the outcomes that display two initial $A$'s to display a $B$ in the third position. Hence the probability for an outcome to begin with the sequence $AAB$ is $(1/3)^2(2/3)^1$. Finally and by extension, the probability for the outcome $AABBA$ is found to be $(1/3)^3(2/3)^2$, as desired. Furthermore, since the number of sequences of length $n$ that display $k$ times an $A$ and $n-k$ times a $B$ is $\binom{n}{k}$, it follows that the probability of the event $E_{n,k}$ that consists precisely of these sequences is

$$P(E_{n,k}) = \binom{n}{k}\left(\frac{1}{3}\right)^k\left(\frac{2}{3}\right)^{n-k}. \tag{1.6}$$

**1.3.15 Example.** To generalize the result of the preceding example, let us assume that a *biased* coin is tossed $n$ times and that the probability for getting a head is $p$. Then, the probability for getting a tail is $1-p$, and, in employing essentially the same argument as in the derivation of (1.6), it follows that for any $k \in \{0,\dots,n\}$, the probability of the event $E_{n,k}$ that consists of all sequences that display $k$ times an $A$ and $n-k$ times a $B$ is the *binomial probability*

$$\boxed{P(E_{n,k}) = \binom{n}{k}p^k(1-p)^{n-k}.} \tag{1.7}$$

**1.3.16 Example.** Assume that a probability experiment generates two outcomes $A$ and $B$ with respective probabilities $p, 1-p \in (0,1)$. The experiment is performed until it results in the outcome $A$ for the first time, and the number of trials that it took to produce this outcome $A$ is recorded. Given this experimental procedure, it follows that the sample space consists of all positive integers, that is, $S = \{1,2,3,\dots\} = \mathbb{N}$ and that the probability of any outcome $k \in S$ is

$$\boxed{P(k) = P(\{k\}) = (1-p)^{k-1}p.} \tag{1.8}$$

This is so because the outcome $k$ occurs precisely in the case where the first $k-1$ trials produce a $B$ and the $k$-th trial produces an $A$. Furthermore and by implication, the probability of $E_n := \{1,2,3,\dots,n\}$ is

$$P(E_n) = \sum_{k=1}^{n}(1-p)^{k-1}p = 1 - (1-p)^n \quad \text{(see the remark below).} \tag{1.9}$$

Note: the limiting probability of $E_n$ as $n$ tends to $\infty$ is 1—as might be expected—because

$$\lim_{n \to \infty} P(E_n) = 1 - \lim_{n \to \infty} (1-p)^n = 1 - 0 = 1.$$

*Remark.* Equation (1.9) is derived from the following formula:

$$\sum_{k=0}^{n} q^k = \frac{1 - q^{n+1}}{1 - q} \qquad (1.10)$$

for all $q \in \mathbb{R} \setminus \{1\}$. For using this formula with $1-p$ in place of $q$, we find that

$$\sum_{k=1}^{n} (1-p)^{k-1} p = p \sum_{k=0}^{n-1} (1-p)^k = \frac{p(1-(1-p)^n)}{1-(1-p)} = 1 - (1-p)^n.$$

Furthermore, in order to derive (1.10), we argue as follows:

$$\sum_{k=0}^{n} q^k = \frac{1-q}{1-q} \sum_{k=0}^{n} q^k = \frac{1}{1-q} \left( \sum_{k=0}^{n} q^k - \sum_{k=0}^{n} q^{k+1} \right)$$

$$= \frac{1}{1-q} \left( \sum_{k=0}^{n} q^k - \sum_{k=1}^{n+1} q^k \right) = \frac{q^0 - q^{n+1}}{1-q} = \frac{1 - q^{n+1}}{1-q}.$$

For later reference we also note that in the case where $q \in (-1, 1)$ we have $\lim_{n \to \infty} q^n = 0$, and therefore,

$$\sum_{k=0}^{\infty} q^k = \lim_{n \to \infty} \sum_{k=0}^{n} q^k = \frac{1 - \lim_{n \to \infty} q^{n+1}}{1-q} = \frac{1}{1-q}. \qquad (1.11)$$

**1.3.17 Example.** Assume that a random experiment is performed by tossing a biased coin (with $P(H) = p$ and $P(T) = 1 - p$) repeatedly until $r$ heads have been recorded (where $r$ is a fixed positive integer). The outcome of the experiment, denoted by $k$, is the number of times that the coin has to be tossed until this quota of $r$ heads has been met. Given this description, it follows that the sample space is $S = \{r, r+1, r+2, \ldots\}$. Furthermore, since the number of sequences in the letters $H$ and $T$ of length $k$ that contain exactly $r$ times an $H$ and also end with an $H$ is equal to $\binom{k-1}{r-1}$ (because the $k$-th letter is fixed and known to be $H$), it follows that the probability of any outcome $k \in S$ is

$$P(k) = P(\{k\}) = \binom{k-1}{r-1} (1-p)^{k-r} p^r.$$

**1.3.18 Example.** A random sample of size $n$ is drawn from a bowl that contains $N$ distinct objects, $r$ of which have a certain trait, and the number of

objects in the sample that have the trait is denoted by $k$. Then the least value
that $k$ can assume is $\max\{0, n - (N - r)\}$ because on the one hand $k$ cannot
be less than zero (this is trivial), and on the other hand, if the sample size $n$
is greater than the number of objects without the trait—which is $N - r$—then
$n - (N - r)$ is positive and is the least possible number of objects in the sample
that have the trait. Similarly, the largest value that $k$ can assume is $\min\{n, r\}$
because clearly, $k$ cannot be larger than either the size of the sample or the
total number of objects that have the trait. Consequently, if we consider $k$ to
be the outcome that is recorded whenever a sample has been drawn, then the
sample space is

$$S = \{\max\{0, n - (N - r)\}, \ldots, \min\{n, r\}\} \subset \mathbb{N} \cup \{0\}.$$

In order to determine the probability $P(k)$, we observe to begin with that, in
general, choosing a sample of size $m$ from a set of $M$ distinct objects is like
attaching a label 'chosen' or 'not chosen'—1 or 0—to each of the $M$ given
objects. That is to say, every ordered sequence of length $M$ in the digits 1 and
0 that contains $m$ times a 1 and $M - m$ times a 0 corresponds to exactly one
possible sample or subset, and since we know the number of these sequences
to be equal to $\binom{M}{m}$, it follows that the number of the samples here in question
is equal to $\binom{M}{m}$ as well. By implication, the total number of samples of size
$n$ that can be drawn from $N$ distinct objects is $\binom{N}{n}$, and the total number of
samples of size $n$ that contain $k$ objects with the given trait and $n - k$ objects
without it is the product $\binom{r}{k}\binom{N-r}{n-k}$ because the first factor in this product is
equal to the number of ways in which $k$ objects with the trait can be drawn
from a total of $r$ objects with the trait, and the second, by analogy, is equal
to the number of ways in which $n - k$ objects without the trait can be drawn
from a total of $N - r$ objects without the trait. Thus we find that

$$P(k) = P(\{k\}) = \frac{\dbinom{r}{k}\dbinom{N - r}{n - k}}{\dbinom{N}{n}}$$

for all $k \in S$.

At first sight it may be tempting to think that in choosing $n$ objects that
do or do not have a trait we are simply dealing with a binomial random process
and that therefore formula (1.7) ought to be applicable. But here we need to be
careful because the probabilities for picking objects with or without the trait
are changing with each pick. For the first pick the probabilities in question
are $r/N$ and $(N - r)/N$, respectively, but for the second pick they are either
$(r - 1)/(N - 1)$ and $(N - r)/(N - 1)$ or $r/(N - 1)$ and $(N - r - 1)/(N - 1)$
depending on whether the first object chosen did or did not have the trait.
However, if we do take these changes in the probabilities into account, then we

do arrive at the same formula as above. Here is how it works: the probability for choosing $k$ times an object with the trait and $n-k$ times an object without the trait in one particular order is always (i.e., regardless of the order) equal to

$$\frac{r(r-1)\cdots(r-k+1)(N-r)(N-r-1)\cdots(N-r-n+k+1)}{N(N-1)\cdots(N-n+1)}$$
$$= \frac{r!(N-r)!(N-n)!}{N!(r-k)!(N-r-n+k)!}.$$

Since the number of ways in which $k$ objects with the trait and $n-k$ objects without the trait can be ordered is $\binom{n}{k}$, it follows that

$$P(k) = \frac{r!(N-r)!(N-n)!}{N!(r-k)!(N-r-n+k)!}\binom{n}{k} = \frac{\binom{r}{k}\binom{N-r}{n-k}}{\binom{N}{n}},$$

as desired.

All the sample spaces that we encountered so far were discrete in that they consisted of either finitely many outcomes or infinitely many outcomes that could be listed one by one in an infinite sequence (as, for instance, the outcomes in the sample space $S = \{1, 2, 3, \ldots\} = \mathbb{N}$ in Example 1.3.16). This is not surprising because continuous sets, such as intervals in $\mathbb{R}$ or filled regions in the two and three-dimensional spaces $\mathbb{R}^2 = \{(x,y) \mid x, y \in \mathbb{R}\}$ and $\mathbb{R}^3 = \{(x,y,z) \mid x, y, z \in \mathbb{R}\}$ can never be encountered in any kind of concrete, empirical reality. After all, measurements are always finite in number and can never be infinitely precise. Consequently, the question of whether the spatio-temporal reality that we commonly believe to inhabit is discrete or continuous in nature is ultimately not a question of fact but rather of personal faith. All the same, however, it can be very helpful to work with sample spaces that are continuous because it is frequently simpler and therefore more practical. If we wish to describe, for example, the probabilistic behavior of a computer that generates random values between 0 and 1 with a 10-digit precision, we certainly can construe the sample space to consist of all the 10-digit decimal values between 0 and 1. But for simplicity's sake we can just as well construe $S$ to be the continuous interval $[0, 1]$. The latter choice is clearly an idealization, but that's okay because it works. In fact, it probably works better by far than a discrete cumbersome set that contains $10^{10}$ decimal values.

However, there also is a problem here. For if $S$ contains, say, $10^{10}$ outcomes that are all equally likely, then each individual outcome $s \in S$ occurs with probability $P(s) = 1/\#S = 10^{-10}$. By contrast, if $S = [0, 1]$, then the requirement that all outcomes be equally likely actually implies that $P(s) = 0$

for all $s \in S = [0, 1]$. After all, if $P(s)$ were positive, then the total probability would be $P(S) = \#S \cdot P(s) = \infty \cdot P(s) = \infty$ which is impossible because $P(S)$ must always be equal to one. So this seems paradoxical because if we imagine a random number generator that produces with each run a number from the interval $[0, 1]$, then each run produces a number that is impossible to occur because its probability is actually zero. Fortunately, though, there is a simple way to make sense of this. For instead of considering events consisting of single points, we may, for example, consider events that themselves are intervals. For instance, if $S = [a, b]$ for some $a, b \in \mathbb{R}$ with $a < b$, then the requirement that all outcomes in $S$ be equally likely is naturally expressed in the demand that for all $c, d \in [a, b]$ with $c < d$ the probability of the event $E = [c, d]$ be equal to the length of the interval $[c, d]$ relative to the length of the sample space $S = [a, b]$, that is,

$$P(E) = \frac{\text{length of } E}{\text{length of } S} = \frac{d - c}{b - a}.$$

This equation clearly is intuitively compelling because if, for example, the length of $E$ is one third of the length of $S$, and if all outcomes in $S$ are equally probable, then surely the probability of $E$ ought to be one third. Similarly, if $S$ is a region or volume in $\mathbb{R}^2$ or $\mathbb{R}^3$ and if all outcomes in $S$ are equally likely to occur, then the probability of any subregion—or event—$E \subset S$ ought to be equal to the relative area or volume content of $E$. That is to say, it ought to be the case that

$$P(E) = \frac{\text{area or volume of } E}{\text{area or volume of } S}.$$

**1.3.19 Example.** A straight horizontal line is drawn through the center of a circle of radius 1 and a point $P$ on the circle is randomly chosen (Figure 1.4). What is the probability that the distance $d$ of $P$ from the horizontal line is

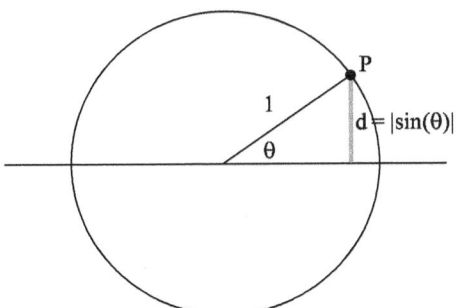

Figure 1.4: a random point on a circle.

greater than one half? To answer this question, we observe to begin with that a random choice of $P$ is equivalent to a random choice of the angle $\theta$ between the given horizontal line and the line that connects $P$ with the center of the circle. Thus the sample space in this example is the set of all angles $\theta$ between 0 and $2\pi$, that is, $S = [0, 2\pi]$. Furthermore, since the distance $d$ is equal to $|\sin(\theta)|$ and since the latter value is greater than $1/2$ for all $\theta \in E := (\pi/6, 5\pi/6) \cup (7\pi/6, 11\pi/6)$, we may infer that

$$P(E) = \frac{(5\pi/6 - \pi/6) + (11\pi/6 - 7\pi6)}{2\pi} = \frac{4\pi/3}{2\pi} = \frac{2}{3}.$$

**1.3.20 Example.** Given that a computer produces at random two numbers $s$ and $t$ between 0 and 4 (i.e., $s, t \in [0, 4]$), we wish to answer the following questions: what is the probability for...

  a) $s$ to be less than or equal to 1 and for $t$ to be greater than or equal to 2?

  b) $s$ to be equal to 1 and for $t$ to be equal to 2?

  c) $s^2$ to be less than or equal to $t$?

  d) $s^2$ to be equal to $t$?

  e) $|s - t|$ to be less than or equal to $1/2$?

The first problem that we face in trying to answer these questions is to determine the sample space. At first sight it may be tempting to think that $S$ is the interval $[0, 4]$, but since we assumed that the computer produces *two* values $s, t \in [0, 4]$, it is actually more adequate to construe $S$ to be the filled square $[0, 4] \times [0, 4] = \{(s, t) \mid s, t \in [0, 4]\} \subset \mathbb{R}^2$. Consequently, the probabilities of the events described in a)-e) will have to be determined as area contents divided by the total area content of $S$ which is $4^2 = 16$.

**a)** If $s \in [0, 1]$ and $t \in [2, 4]$, then $E = [0, 1] \times [2, 4]$ (Figure 1.5), and therefore,

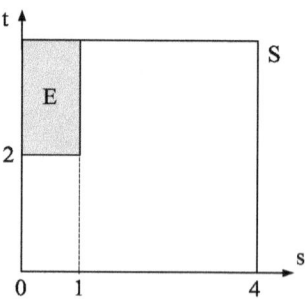

Figure 1.5: the event $E = [0, 1] \times [2, 4]$.

$P(E) = (1 - 0) \cdot (4 - 2)/16 = 1/8$.
**b)** If $s = 1$ and $t = 2$, then $E = \{(1, 2)\}$, and since a single point has area zero, it follows that $P(E) = 0$.
**c)** Here we have $E = \{(s, t) \in [0, 4] \times [0, 4] \mid s^2 \leq t\}$ (Figure 1.6), and therefore,
$P(E) = \left(8 - \int_0^2 s^2 \, ds\right)/16 = (8 - 2^3/3)/16 = 1/3$.

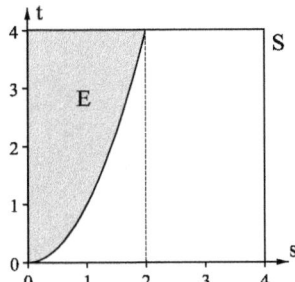

Figure 1.6: the event $E = \{(s,t) \in S \mid s^2 \leq t\}$.

**d)** If $E = \{(s,t) \in [0,4] \times [0,4] \mid s^2 = t\}$, then $E$ is a segment of a curve—a parabola—and as such has vanishing area content. Hence $P(E) = 0$.

**e)** If $E = \{(s,t) \in [0,4] \times [0,4] \mid |s-t| \leq 1/2\}$, then $P(E)$ is the relative area of the region shown in Figure 1.7, and therefore, $P(E) = (16 - (7/2)^2)/16 = 15/16$.

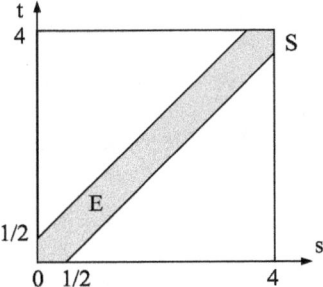

Figure 1.7: the event $E = \{(s,t) \in S \mid |s-t| \leq 1/2\}$.

**1.3.21 Example.** We break a stick in two places. What is the probability that the length of each of the three pieces thus obtained is greater than or equal to one fourth of the length of the stick? Since the answer to this question evidently does not depend on the length of the stick, we will assume for convenience that its length is equal to one. Consequently, the stick may be identified with the interval $[0,1]$, and the two random places at which it is broken may be construed to be two random numbers $s, t \in [0,1]$. Thus the sample space $S$ is the unit square $[0,1] \times [0,1] = \{(s,t) \mid s,t \in [0,1]\}$. Considering first the case where $s \leq t$, we observe that the lengths of the three pieces that the break points $s$ and $t$ generate are $s$, $t-s$, and $1-t$. Consequently, the requirement that the length of each of these pieces be greater than or equal to $1/4$, is properly expressed in the following inequalities:

$$s \geq 1/4,$$

$$t - s \geq 1/4,$$
$$1 - t \geq 1/4,$$

or equivalently,

$$s \geq 1/4,$$
$$t \geq s + 1/4, \hspace{3cm} (1.12)$$
$$3/4 \geq t.$$

The region in $S$ that these three inequalities specify is the right triangle with vertices at $(1/4, 1/2)$, $(1/2, 3/4)$, and $(1/4, 3/4)$. Moreover, in a completely analogous fashion we also find that in the case where $t \leq s$ the corresponding inequalities are

$$t \geq 1/4,$$
$$s \geq t + 1/4, \hspace{3cm} (1.13)$$
$$3/4 \geq s,$$

and the corresponding region in $S$ is the right triangle with vertices at $(1/2, 1/4)$, $(3/4, 1/2)$, and $(3/4, 1/4)$. By implication, the event $E$ that consists of all points $(s, t) \in S$ that generate three line segments that are each greater than or equal to $1/4$ in length is equal to the union of the two triangular regions shown in Figure 1.8. Hence $P(E) = (1/4)^2 = 1/16$.

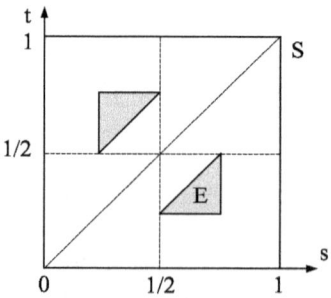

Figure 1.8: the event $E$.

To further explore this present example, we replace the value $1/4$ with a randomly chosen value $x \in [0, 1]$ and ask the following analogous question: what is the probability for the three pieces that the stick is broken into to be each greater than or equal to $x$ in length? In other words, the question here is how likely it is for us to pick a point $(s, t, x)$ in the extended sample space $S = [0, 1] \times [0, 1] \times [0, 1]$ such that—in analogy to (1.12) and (1.13)—it is the case that either

$$s \geq x,$$
$$t \geq s + x,$$
$$1 - x \geq t$$

or

$$t \geq x,$$
$$s \geq t + x,$$
$$1 - x \geq s.$$

Since the length of the stick is one, it follows that the three pieces that the stick is broken into cannot be each longer than $1/3$. Consequently, the set

$$E_x := \{(s,t) \in [0,1] \times [0,1] \mid s \geq x \wedge t \geq s + x \wedge 1 - x \geq t\}$$
$$\cup \{(s,t) \in [0,1] \times [0,1] \mid t \geq x \wedge s \geq t + x \wedge 1 - x \geq s\}.$$

is empty whenever $x > 1/3$. Furthermore, if $x \in [0, 1/3]$, then $E_x$ is easily seen to be the union of two right triangles in $[0,1] \times [0,1]$ with vertices at $(x, 2x)$, $(1 - 2x, 1 - x)$, $(x, 1 - x)$ and $(2x, x)$, $(1 - x, 1 - 2x)$, $(1 - x, x)$, respectively (Figure 1.9). Hence the area of $E_x$ is

$$A(x) = (1 - 3x)^2,$$

and therefore, the probability—or relative volume—of the event $E$ here in question is

$$P(E) = P\left(\bigcup_{x \in [0,1/3]} E_x \times \{x\}\right) = \frac{\int_0^{1/3} A(x)\, dx}{\text{volume of } S} = \frac{-\frac{1}{9}(1 - 3x)^3 \Big|_0^{1/3}}{1} = \frac{1}{9}.$$

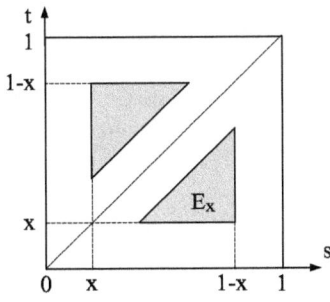

Figure 1.9: the set $E_x$ in $[0,1] \times [0,1]$.

**1.3.22 Example.** (The Buffon Needle) Assume that a needle of length $L$ is tossed at random onto an infinitely extended plane on which infinitely many parallel lines have been drawn in such a way that the distance between any two neighboring lines is always equal to one. We wish to find the probability for the needle to *not* cross any of the drawn lines. To do so, we first need to identify the sample space: each time the needle is tossed, its midpoint position between two lines will be at a distance $x \in [0,1]$ from the lower line and its upper half will be slanted at an angle $\theta \in [0,\pi]$ away from the positive horizontal direction (Figure 1.10). Consequently, the sample space $S$ may properly be chosen to be

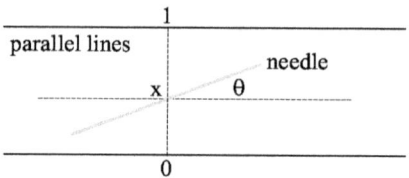

Figure 1.10: position of a needle.

the rectangle $[0,\pi] \times [0,1] = \{(\theta, x) \mid \theta \in [0,\pi] \wedge x \in [0,1]\}$. Furthermore, since the vertical coordinates of the two ends of the needle are $x \pm (L/2)\sin(\theta)$, we may infer that the event $E$ of 'not crossing' occurs if and only if the following two inequalities are satisfied:

$$x + (L/2)\sin(\theta) < 1,$$
$$x - (L/2)\sin(\theta) > 0,$$

or equivalently,

$$x < 1 - (L/2)\sin(\theta),$$
$$x > (L/2)\sin(\theta).$$

In order to identify the set $E \subset S$ that is described by these two inequalities, we need to distinguish two cases.

*Case 1: $L < 1$.*
In this case the curves described by the equations $x = (L/2)\sin(\theta)$ and $x = 1-(L/2)\sin(\theta)$ do not intersect, and the event $E$ is the region enclosed between them (Figure 1.11). Using integration, we find that

$$P(E) = \frac{\pi - 2(L/2)\int_0^\pi \sin(\theta)\, d\theta}{\pi} = 1 - \frac{2L}{\pi}. \tag{1.14}$$

*Case 2: $L \geq 1$.*
Here the curves described by the equations $x = (L/2)\sin(\theta)$ and $x = 1 -$

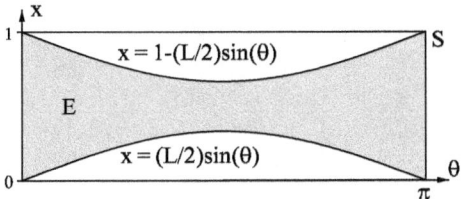

Figure 1.11: the event $E$ in case that $L \leq 1$.

$(L/2) \sin(\theta)$ intersect at the points in $S$ whose $\theta$-coordinates satisfy the equation

$$(L/2) \sin(\theta) = 1 - (L/2) \sin(\theta),$$

or equivalently,

$$\sin(\theta) = \frac{1}{L}.$$

Since the solutions of the latter equation in $[0, \pi]$ are $\theta = \arcsin(1/L)$ and $\theta = \pi - \arcsin(1/L)$, it follows that the event $E$ is the union of the shaded regions in Figure 1.12. Since these regions are evidently equal in size, we may

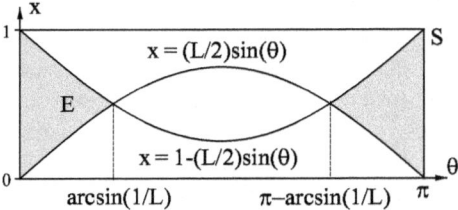

Figure 1.12: the event $E$ in case that $L > 1$.

conclude that

$$P(E) = \frac{2 \left( \arcsin(1/L) - 2(L/2) \int_0^{\arcsin(1/L)} \sin(\theta)\, d\theta \right)}{\pi}$$

$$= \frac{2 \left( \arcsin(1/L) - L(-\cos(\arcsin(1/L)) + 1) \right)}{\pi}$$

$$= \frac{2 \left( \arcsin(1/L) + L\sqrt{1 - (1/L)^2} - L \right)}{\pi}$$

$$= \frac{2}{\pi} \left( \arcsin(1/L) + \sqrt{L^2 - 1} - L \right).$$

*Remark.* The experiment discussed in the preceding example was proposed by the noted French naturalist Georges-Louis Leclerc, Comte de Buffon (1707–1788), as a so-called Monte-Carlo method for determining the value of $\pi$. For as

we consider the complementary event of crossing—rather than not crossing—
in the first case where $L < 1$, we notice that, according to equation (1.14), the
corresponding probability is

$$1 - P(E) = \frac{2L}{\pi},$$

and in particular, for $L = 1/2$, the probability is $1/\pi$. So as we toss a needle of
length $1/2$ repeatedly onto a board on which parallel lines have been drawn at
the constant distance 1, and as we count the number of times that the needle
crosses one of these lines, we can expect to find that

$$\frac{\text{number of crossings}}{\text{number of tosses}} \approx \frac{1}{\pi}$$

or, equivalently,

$$\boxed{\pi \approx \frac{\text{number of tosses}}{\text{number of crossings}}.}$$

**1.3.23 Example.** (Bertrand's Paradox) Assume that a chord is drawn at
random inside a circle of radius $R$. What is the probability that the length of
the chord will be greater than the side length of an equilateral triangle inscribed
into the same circle? Surprisingly, the answer to this question depends on what
we mean by 'drawing a chord at random'. One way to interpret this phrase is
to say that we mark at random two points on a given circle and then connect
them with a line. In that case it clearly doesn't matter where we place the
first point—what matters is only the position of the second point *relative* to
the first. Furthermore and in the light of the two cases shown in Figure 1.13, it

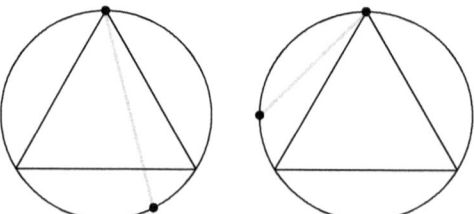

Figure 1.13: Bertrand's Paradox, first case.

is clear that the length of the chord will be greater than the side length of the
inscribed equilateral triangle if the second point is placed in that third of the
circle that is opposite to the first point. Consequently, the probability of the
event $E$ here in question—the chord is longer than the side of the triangle—is
$P(E) = 1/3$.

So far so good, but there is an alternative. For since the perpendicular
bisector of every chord that we draw passes through the circle's center, we

can also pick a random direction—an angle between 0 and $2\pi$—draw a radial line from the center in that direction, pick a point $x$ at random on that radial line, and draw a chord through $x$ at a right angle relative to the radial line (Figure 1.14). By symmetry, the choice of the direction is completely irrelevant,

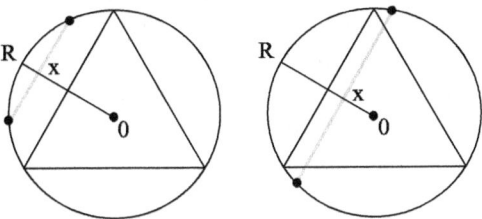

Figure 1.14: Bertrand's Paradox, second case.

and the random process that effectively matters is therefore only the picking of a point $x$ from the radial line, or equivalently, from the interval $[0, R]$. Since elementary geometry readily shows that the length of the perpendicular chord is greater than the side length of the inscribed triangle whenever $x < R/2$, it follows that $P(E) = (R/2)/R = 1/2$.

Finally, what we also can do is to pick a point $P$ at random from the disc that the circle encloses, draw a radial line through $P$, and then a chord, also through $P$, that forms a right angle with the radial line (Figure 1.15). Proceeding in this way, we find that the chord is longer than the sides of

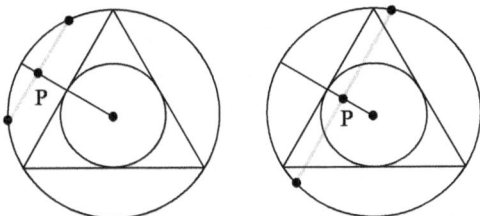

Figure 1.15: Bertrand's Paradox, third case.

the inscribed triangle if and only if $P$ is in the triangle's inner circle. Since the radius of this inner circle is $R/2$, we may conclude that in this case the probability of $E$ is $P(E) = \pi(R/2)^2/(\pi R^2) = 1/4$.

So in summary, the probability in question is either 1/3, 1/2, or 1/4 depending on the mode of interpretation that we happen to favor. Naturally, this conclusion may seem rather perplexing, but upon closer inspection the paradox dissolves quite readily. For why would we expect three completely different and entirely disconnected random processes to yield equal probabilities for mutually unrelated events? Picking a point at random from a circle is one random event, choosing a point from an interval is another, and selecting a point from inside a disc is yet another. Seen in this light, the drawing of the chord is an inessential add-on that serves no purpose other than to hide the fact that the underlying random processes are indeed entirely different. In the first case, the event $E$ is a segment on a circle, in the second it is half an interval—and should be denoted by a different letter like $F$—and in the third it is a disc inside a disc, which should perhaps be labeled $D$. Put differently, the instruction to draw at random a chord in a circle is meaningless as long as this instruction is not properly defined. One definition yields one result and another yields another, and hence there really is no paradox.

# Exercises

**1.3.24.** Prove equation (1.4).

**1.3.25.** Show that the two formulas for $P(E)$, derived in Example 1.3.22, yield the same value in the boundary case $L = 1$.

**1.3.26.** Assume that a sample of 500 randomly selected passenger cars contains 40 defective vehicles. Given this information, what is your best estimate for the probability that a randomly selected passenger car is defective?

**1.3.27.** An engine is tested at four different engine speeds and five different engine temperatures. Given this information, what is the total number of test settings?

**1.3.28.**   a) What is the number of 01-strings of length five (i.e., sequences of length five that are composed only of zeros and ones)?

  b) What is the number of 01-strings of length $n$?

  c) What is the number of subsets of a finite set $S$ if $\#S = n$?

**1.3.29.** A shipment of thirty watches contains four watches that are defective. What is the probability that a random quality-control sample of six watches (drawn from the thirty watches) does not contain any watches that are defective?

**1.3.30.**   a) What is the number of words of length 9 that contain no letters other than $a$, $b$, and $c$?

  b) What is the probability that a randomly selected word of length 9 in the letters $a$, $b$, and $c$ contains exactly three times the letter $a$, twice the letter $b$, and four times the letter $c$?

**1.3.31.** Consider the number 2,235,671. How many decimal numbers of length 7 that do not contain a zero differ from this given number in exactly two digits?

**1.3.32.** A bowl contains 5 blue balls and 3 red ones. If two balls are drawn at random and if the first ball that is drawn is *not* returned to the bowl before the second ball is drawn, then what is the probability that one of these two balls is red and the other one blue? What is the probability that this outcome occurs if the first ball *is* returned to the bowl before the second one is drawn?

**1.3.33.** A probability experiment is performed by rolling two dice and multiplying the numbers that the dice happen to show. In other words, the outcome of the experiment is a single number obtained by multiplying two given numbers. Find the sample space of this experiment and determine the probability of each element in the sample space.

**1.3.34.** What are the probabilities of each of the elements in the sample spaces described in Exercises 1.2.11 and 1.2.12?

**1.3.35.** A pair of dice—one black, one red—is rolled once and the difference of the numbers—black minus red—is recorded. What is the sample space of this experiment and what are the probabilities of the outcomes in the sample space?

**1.3.36.** Three dice are rolled once and the three numbers that they show are recorded. What is the probability of the following event: the first die shows a one, the second a 2, 3, or 4, and the third shows a number different from 6?

**1.3.37.** A board is ruled into unit squares and a disc of radius $a < 1$ is tossed at random. What is the probability for the disc not to cross any of the drawn lines? Plot this probability as a function of $a$. For which value of $a$ is the probability equal to $1/2$?

**1.3.38.** Assume that you walk a random distance $x \in [0, 1]$ away from a straight road at a random angle $\theta \in [0, \pi]$. What is the probability that your distance from the road at the end of your walk is greater than $1/2$?

**1.3.39.** If you select two random numbers in the interval $[0, 2]$, then what is the probability that the product of these numbers is less than $1/2$?

**1.3.40.** A random number $x$ is drawn from the interval from 0 to 7. If $A$ is the area of the circle the circumference of which is $x$, then what is the probability for $A$ to be less than $9/\pi$?

# 1.4 The Axioms of Probability

In several of the preceding examples we saw that the calculation of a probability can involve an integration. This is no coincidence because, very broadly speaking, the theory of probability is nothing but the theory of integration plus the added concept of *independence* (which will be defined below, in Section 1.5). Indeed, if the notion of an integral is properly generalized, then all probabilities can be construed to be values of integrals, and all sample spaces can be viewed as sets that are endowed with a structure that makes possible the definition of precisely these integrals. In taking this approach the theory of probability can be given a maximally general conceptual foundation, but it also becomes fairly abstract. Thus in order to avoid this higher level of abstraction, we will adopt in the present, more elementary exposition a down-to-earth practical approach that's easier to grasp. We will not discuss how integrals in general are defined and how sample spaces need to be structured in order to make integration possible, and we will also not attempt to describe exactly how this structuring imposes a restriction on the notion of an event. For strictly speaking, an event is not just a subset of a sample space but rather a permissible subset that satisfies certain defining properties. So the notion of a sample space and the attendant notion of an event will be left imprecise throughout this text, and the notion of an integral will not be encountered outside the context of ordinary multivariable calculus. However, what we can and will do is to identify what basic properties a measure of probability $P$ must satisfy given the assumption that the set of events in a certain sample space has somehow been adequately defined. Hence we introduce the following definition:

**1.4.1 Definition.** Two events $E$ and $F$ in a sample space $S$ are said to be *mutually exclusive* if $E \cap F = \emptyset$. Furthermore, given an index set $I$, we say that $(E_i)_{i \in I}$ is *a collection—or sequence—of mutually exclusive events* if each $E_i$ is an event in $S$ and if $E_i \cap E_j = \emptyset$ whenever $i \neq j$.

**1.4.2 Example.** If $I := \{1, 2, 3, \ldots\} = \mathbb{N}$, $S := [0, 1]$, and

$$E_i := \left( \frac{1}{i+1}, \frac{1}{i} \right],$$

then $(E_i)_{i \in I} = (E_i)_{i \in \mathbb{N}}$ is a collection of mutually exclusive events because

$$\left( \frac{1}{i+1}, \frac{1}{i} \right] \cap \left( \frac{1}{j+1}, \frac{1}{j} \right] = \emptyset$$

whenever $i \neq j$.

**The Axioms of Probability.** Assume that $S$ is a sample space and that a suitable set of events $E \subset S$ has been defined so that, in particular, the empty set $\emptyset$ and the whole set $S$ are included among these events. Then a probability measure $P$ is a function that assigns to every event $E \subset S$ a real number $P(E)$ in such a way that the following properties are satisfied:

(i) $P(S) = 1$,

(ii) $P(E) \geq 0$ for all events $E \subset S$,

(iii) $P(E_1 \cup E_2 \cup E_3 \cup \ldots) = P(E_1) + P(E_2) + P(E_3) + \ldots$ for any finite or infinite collection of mutually exclusive events $E_1, E_2, E_3, \ldots \subset S$.

**1.4.3 Theorem.** *The following properties are satisfied by any probability measure $P$:*

a) $P(\emptyset) = 0$,

b) $P(S \setminus E) = 1 - P(E)$ for all events $E \subset S$,

c) $P(E) \leq P(F)$ whenever $E \subset F$,

d) $P(E) \leq 1$ for all events $E \subset S$,

e) $P(E \cup F) = P(E) + P(F) - P(E \cap F)$ for all events $E, F \subset S$.

*Note: the fact that $S \setminus E$, $E \cup F$, and $E \cap F$ are events whenever $E$ and $F$ are events follows from the suitability condition in the axioms of probability and will here be accepted without proof.*

*Proof.* We will prove a), b), and e) and leave the proofs of c) and d) as exercises to the reader.

**a)** Since $\emptyset \cap \emptyset = \emptyset$, axiom (iii) implies that

$$P(\emptyset) = P(\emptyset \cup \emptyset) = P(\emptyset) + P(\emptyset),$$

and in subtracting $P(\emptyset)$ from both sides of this equation, we find that $P(\emptyset) = 0$, as desired.

**b)** Since $E \cap (S \setminus E) = \emptyset$, we may apply axioms (i) and (iii) to conclude that

$$1 = P(S) = P(E \cup (S \setminus E)) = P(E) + P(S \setminus E),$$

and therefore, $P(S \setminus E) = 1 - P(E)$.

**e)** Since $E$ is a disjoint union of $E \setminus F$ and $E \cap F$, axiom (iii) implies that

$$P(E) = P(E \setminus F) + P(E \cap F),$$

and similarly we also find that

$$P(F) = P(F \setminus E) + P(E \cap F).$$

Furthermore, since $E \cup F$ is a disjoint union of $E \setminus F$, $E \cap F$, and $F \setminus E$, axiom (iii) further implies that

$$\begin{aligned}
P(E \cup F) &= P(E \setminus F) + P(E \cap F) + P(F \setminus E) \\
&= P(E) - P(E \cap F) + P(E \cap F) + P(F) - P(E \cap F) \\
&= P(E) + P(F) - P(E \cap F),
\end{aligned}$$

as desired. $\square$

# Exercises

**1.4.4.** Prove Theorem 1.4.3c,d.

**1.4.5.** The probability for a newly produced light bulb to be defective is $1/500$. What is the probability for it to be non-defective? Which axiom/fact discussed in this section is relevant to this problem?

**1.4.6.** Assume that $E$ and $F$ are events in a sample space $S$ such that $P(E) = 0.6$, $P(F) = 0.7$, and $P(S \setminus (E \cup F)) = 0.1$. Use this information to determine $P(E \cap F)$.

**1.4.7.** If $F$ and $G$ are events in a sample space $S$, then $P(E \cup F) = P(E) + P(F) - P(E \cap F)$. Generalize this result to the case of three events by completing the equation $P(E \cup F \cup G) = P(E) + P(F) + P(G) - \ldots$

**1.4.8.** Find $P((E \cup F \cup G) \setminus ((E \cap F) \cup (E \cap G) \cup (F \cap G)))$ under the assumption that $P(E \cup F \cup G) = 0.8$, $P(E \cap F \cap G) = 0.1$, and $P(E \cap F) = P(E \cap G) = P(F \cap G) = 0.2$.

**1.4.9.** Assume that $S$ is a sample space and that $E$, $F$, and $G$ are events in $S$ that satisfy the following conditions: $P((E \cap F) \cup (E \cap G)) = P(F \cup G)$ and $P(E \cup F \cup G) = 3/4$. Use this information to determine $P(E)$.

**1.4.10.** Assume that $E$ and $F$ are events in a sample space $S$ such that $P(E \cap F) = 0.5$ and $P(E \cup F) = 0.7$. Given this information, what is the least value that the product $P(E)P(F)$ can assume?

**1.4.11.** Show that $P(E)P(F) \geq P(E \cap F)P(E \cup F)$ for all events $E$ and $F$ in a sample space $S$.

**1.4.12.** Show that

$$\sqrt{P(E)P(F)} \leq \frac{P(E \cap F) + P(E \cup F)}{2}$$

for all events $E$ and $F$ in a sample space $S$.

**1.4.13.** Show that

$$P(E \cap F \cap G) \geq P(E) + P(F) + P(G) - 2$$

for all events $E$, $F$, and $G$ in a sample space $S$.

**1.4.14.** Use Exercise 1.4.13 to explain why $P(E \cap F \cap G)$ must be strictly greater than zero whenever $P(E)$, $P(F)$, and $P(G)$ are each strictly greater than $2/3$, and find an example of a sample space $S$ and three events $E, F, G \subset S$ such that $P(E) = P(F) = P(G) = 2/3$ and $P(E \cap F \cap G) = 0$.

## 1.5   Conditional Probability and Independence

As mentioned above, probability at bottom is the study of integration plus independence as an added conception. In other words, independence as a concept is absolutely central. So what do we mean when we say that two events $E$ and $F$ are independent? Well, intuitively and not surprisingly we mean by it that the occurrence of the event $E$ is not in any way conditioned or affected by the simultaneous occurrence of the event $F$. For instance, if we toss a coin three times, so that the sample space is

$$S = \{HHH, HHT, HTH, HTT, THH, THT, TTH, TTT\}, \qquad (1.15)$$

then the event $E$ that the first coin toss produces an $H$ ought to be and is in fact independent of the event $F$ that the second coin toss produces a $T$. After all, a coin doesn't have any memory, and—presumably—the outcome of the first toss can therefore not affect in any way the outcome of the second. In other words,

$$E = \{HHH, HHT, HTH, HTT\} \qquad (1.16)$$

is independent of

$$F = \{HTH, HTT, TTH, TTT\}. \qquad (1.17)$$

Intuitively, this makes perfect sense, but how do we capture this common sense notion of independence in a rigorous mathematical definition? To find the answer, we may argue as follows: if $E$ is independent of $F$, then our knowledge of the occurrence or non-occurrence of $E$ should have no bearing on the parallel occurrence or non-occurrence of $F$. In other words, the probability for $E$ to occur given that $F$ has already occurred should be exactly the same as the probability of $E$. So in denoting by $P(E|F)$ the probability of $E$ given that we know that $F$ has occurred, we may say that $E$ and $F$ are independent if

$$P(E|F) = P(E).$$

Furthermore, in assuming that $F$ occurs, we essentially reduce the sample space from $S$ to $F$ because from the outset we consider only those outcomes that the event $F$ comprises. Consequently, it is natural and adequate to define $P(E|F)$ to be the relative probability of $E$ in $F$, or equivalently, the probability of $E \cap F$ divided by the probability of $F$.

**1.5.1 Definition.** The *conditional probability* for an outcome of a random experiment to be an element of an event $E$ given that it is known to be an element of an event $F$ is

$$P(E|F) := \frac{P(E \cap F)}{P(F)}.$$

**1.5.2 Definition.** Two events $E$ and $F$ are said to be *independent* if

$$P(E|F) = P(E),$$

or equivalently, if

$$P(E \cap F) = P(E)P(F).$$

Let's see how this latter definition applies to the events $E$ and $F$ in (1.16) and (1.17), respectively: the intersection of $E$ and $F$ is

$$E \cap F = \{HTH, HTT\}.$$

By implication, if the coin we use is a biased coin with probability $p$ for $H$ and $1 - p$ for $T$, then

$$
\begin{aligned}
P(E) &= p^3 + p^2(1-p) + p^2(1-p) + p(1-p)^2 \\
&= p(p^2 + 2p(1-p) + (1-p)^2) = p(p + (1-p))^2 = p, \\
P(F) &= p^2(1-p) + p(1-p)^2 + p(1-p)^2 + (1-p)^3 \\
&= (1-p)(p^2 + 2p(1-p) + (1-p)^2) = 1 - p,
\end{aligned}
\tag{1.18}
$$

and

$$P(E \cap F) = p^2(1-p) + p(1-p)^2 = p(1-p)(p + (1-p)) = p(1-p).$$

Hence $E$ and $F$ are independent, according to Definition 1.5.2, because

$$P(E \cap F) = P(E)P(F). \tag{1.19}$$

For clarity we wish to add that the definition of independence stated in Definition 1.5.2 is purely mathematical in nature and therefore somewhat abstract. So we cannot expect that in merely looking at the defining equation $P(E \cap F) = P(E)P(F)$ we can somehow perceive the physical independence of two successive coin tosses. That would clearly be asking too much. The mathematical definition of independence meaningfully relates to the physical independence of successive coin tosses or rolls of a die, but ultimately, there is nothing concretely physical or temporal about an equation that can be universally used to characterize events as 'independent' in any odd sample space.

Finally, in speaking of "tosses" or "rolls," as we just did, we naturally invoke the possibility that the plural form here signifies an arbitrary quantity greater than one. That is to say, the problem thus suggested is to generalize the notion of independence from a couple of events $E$ and $F$ to an arbitrary finite collection $E_1, E_2, \ldots, E_n$.

**1.5.3 Definition.** Finitely many events $E_1, E_2, \ldots, E_n \subset S$ (with $n > 1$) are said to be *independent* if for all nonempty sets $I \subset \{1, \ldots, n\}$ with more than one element it is the case that

$$P\left(\bigcap_{i \in I} E_i\right) = \prod_{i \in I} P(E_i). \tag{1.20}$$

That's the definition, but what in the world does it mean? Well, consider again the sample space in (1.15) and consider in particular the case where $n$ in Definition 1.5.3 is equal to 3 and where the events $E_1$, $E_2$, and $E_3$ are defined as follows: $E_1$ is the event $E$ displayed in (1.16), $E_2$ is equal to $F$ in (1.17), and $E_3$ is the event that consists of all outcomes that end with a head, that is,

$$E_3 = \{HHH, HTH, THH, TTH\}.$$

Intuitively, the events $E_1$ (first toss produces an $H$), $E_2$ (second toss produces a $T$), and $E_3$ (third toss produces an $H$) ought to be independent, but are they actually independent in the more rigorous sense of Definition 1.5.3? To see why the answer is, "yes," we need to consider all subsets $I$ of $\{1, 2, 3\}$ that contain more than one element. So we need to consider the sets $I = \{1, 2\}$, $\{1, 3\}$, $\{2, 3\}$, and $\{1, 2, 3\}$. For $I = \{1, 2\}$ the defining equation (1.20) assumes the form

$$P(E_1 \cap E_2) = P\left(\bigcap_{i \in \{1,2\}} E_i\right) = \prod_{i \in \{1,2\}} P(E_i) = P(E_1)P(E_2), \qquad (1.21)$$

and similarly, for $I = \{1, 3\}$ and $\{2, 3\}$ the resulting equations are

$$P(E_1 \cap E_3) = P(E_1)P(E_3) \qquad (1.22)$$

and

$$P(E_2 \cap E_3) = P(E_2)P(E_3). \qquad (1.23)$$

Finally, $I = \{1, 2, 3\}$ yields

$$P(E_1 \cap E_2 \cap E_3) = P\left(\bigcap_{i \in \{1,2,3\}} E_i\right) = \prod_{i \in \{1,2,3\}} P(E_i) \qquad (1.24)$$
$$= P(E_1)P(E_2)P(E_3).$$

The validity of (1.21) we already established in (1.19), and the validity of (1.22) and (1.23) can be established very easily in a completely analogous fashion. Concerning the validity of (1.24), we observe that $P(E_1) = P(E) = p$ and $P(E_2) = P(F) = 1 - p$ (by (1.18) above) and that, similarly, $P(E_3) = p$. Consequently, since

$$E_1 \cap E_2 \cap E_3 = \{HTH\},$$

it follows that

$$P(E_1 \cap E_2 \cap E_3) = p(1 - p)p = P(E_1)P(E_2)P(E_3),$$

as desired.

So $E_1$, $E_2$, and $E_3$ are indeed independent in the sense of Definition 1.5.3, but lamentably, the process of establishing independence seems rather cumbersome. For it requires that we check four separate identities—the equations (1.21), (1.22), (1.23), and (1.24). And what is worse, if we increase $n$ from 3 to 4, the number of identities increases from 4 to 11 because 11 is the number of subsets $I$ of $\{1, 2, 3, 4\}$ with more than one element. Naturally, therefore, we may wonder whether there is not some way to reduce the number of identities. Perhaps the validity of (1.21), (1.22), and (1.23) somehow implies the validity of (1.24) or vice versa. To see why the latter suggestion is wrong, we will construct two simple counterexamples. First we consider the case where $S = \{1, 2, 3, 4\}$ and $P(k) = 1/4$ for all $k \in S$. Setting $E_1 := \{1, 2\}$, $E_2 := \{2, 3\}$, and $E_3 := \{2, 4\}$, it follows that

$$P(E_1 \cap E_2) = P(E_1 \cap E_3) = P(E_2 \cap E_3) = P(\{2\}) = \frac{1}{4}$$
$$= \frac{1}{2} \cdot \frac{1}{2} = P(E_1)P(E_2) = P(E_1)P(E_3) = P(E_2)P(E_3),$$

but that nonetheless $E_1$, $E_2$, $E_3$ are not independent because

$$P(E_1 \cap E_2 \cap E_3) = P(\{2\}) = \frac{1}{4} \neq \frac{1}{8} = P(E_1)P(E_2)P(E_3).$$

This shows that the validity of (1.21), (1.22), and (1.23) does not imply the validity of (1.24). And to see why the reverse implication fails as well, we modify our example as follows: the sets $S$, $E_1$, and $E_2$ are left unchanged, but $E_3$ is re-defined to be equal to $E_2 = \{2, 3\}$, and the outcome probabilities are defined to be $P(1) = 3/8$, $P(2) = 1/8$, $P(3) = 3/8$, and $P(4) = 1/8$. Then

$$P(E_1 \cap E_2 \cap E_3) = P(\{2\}) = \frac{1}{8} = \frac{1}{2} \cdot \frac{1}{2} \cdot \frac{1}{2} = P(E_1)P(E_2)P(E_3),$$

but none of the equations (1.21), (1.22), and (1.23) is satisfied because

$$P(E_1 \cap E_2) = P(E_1 \cap E_3) = P(\{2\}) = 1/8$$
$$\neq 1/4 = P(E_1)P(E_2) = P(E_1)P(E_3)$$

and because

$$P(E_2 \cap E_3) = P(E_2) = P(\{2, 3\}) = 1/2 \neq 1/4 = P(E_2)P(E_3).$$

**1.5.4 Example.** If a biased coin with $P(H) = p$ and $P(T) = 1 - p$ is tossed $n$ times, so that $S$ consists of all ordered sequences of length $n$ in the letters $H$ and $T$, and if $E_k$ for any $k \in \{1, \ldots, n\}$ is the event consisting of all outcomes in $S$ that display an $H$ in the $k$-th letter position, then $E_1, \ldots, E_n$ are independent events. This is so, because for all sets $I \subset \{1, \ldots, n\}$ with $\#I > 1$ it is fairly easy to see that

$$P\left(\bigcap_{i \in I} E_i\right) = p^{\#I} = \prod_{i \in I} P(E_i).$$

The purpose of the next example is to provide the reader with a somewhat deeper understanding of the notion of independence by making more immediately apparent how its abstract mathematical description is reflective of our common intuition. This is helpful but not entirely trivial, and readers who find it confusing are therefore kindly advised to move on to Chapter 2. For knowledge of Example 1.5.5 is really not essential to anything that follows.

**1.5.5 Example.** If

$$E_k := \bigcup_{i=1}^{2^{k-1}} \left[\frac{2(i-1)}{2^k}, \frac{2(i-1)+1}{2^k}\right], \tag{1.25}$$

for all positive integers $k$, that is, if

$$E_1 = [0, 1/2],$$
$$E_2 = [0, 1/4] \cup [1/2, 3/4],$$
$$E_3 = [0, 1/8] \cup [1/4, 3/8] \cup [1/2, 5/8] \cup [3/4, 7/8],$$
$$\vdots$$

then, for any positive integer $n$, the events $E_1, \ldots, E_n \subset [0, 1]$ are independent (under the natural additional assumption that the probability of any interval $I \subset [0, 1]$ is defined to be the length of $I$). For instance, for $n = 2$ we have

$$P(E_1 \cap E_2) = P([0, 1/4]) = 1/4 = P(E_1)P(E_2),$$

and for $n = 3$, the validity of (1.21), (1.22), (1.23), and (1.24) is easily established by inspection. Furthermore and in reference to Definition 1.5.3, we find quite readily that for any $n > 1$ and any $I \subset \{1, \ldots, n\}$ with $\#I > 1$ it is the case that

$$P\left(\bigcap_{i \in I} E_i\right) = \frac{1}{2^{\#I}} = \prod_{i \in I} P(E_i).$$

This example is useful because it provides us with a kind of prototypical image of what a collection of independent sets looks like, geometrically speaking. For as we draw the sets $E_1$, $E_2$, and $E_3$, identified above, we obtain the picture shown in Figure 1.16 (which can be easily extended to include sets $E_k$ with indices $k > 3$). So this is the picture that we may want to have in mind as a visual reference point whenever the notion of independence for more than two events is somehow invoked or appealed to.

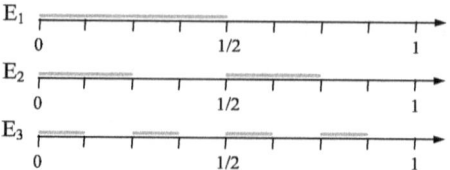

Figure 1.16: prototypical independent events in $[0, 1]$.

That said, however, there still is a problem we need to address. For while the image displayed in Figure 1.16 may be useful, so to speak, as a visual anchor, it doesn't really seem to elucidate the notion of independence insofar as it relates to actual *independent* random trials that are performed one after another without any physical or logical link between them. Where do we perceive in the abstract sets in Figure 1.16 an actual random experiment in actual material reality? On the surface, the answer appears to be, "nowhere," but upon closer inspection, we can in fact discover a very natural link to our standard coin tossing experiment. So let's assume that we toss a perfectly unbiased coin with $P(H) = P(T) = 1/2$ repeatedly, but let's assume as well that this time we keep going indefinitely. That is to say, each outcome in the sample space $S$ that describes this experiment is an *infinite* sequence in the letters $H$ and $T$. Naturally, whether we use the symbols $H$ and $T$ or 0 and 1 doesn't really make much of a difference. So we may as well think of $S$ as consisting of all the infinite sequences in the numbers 0 and 1. Next we also can assign a numerical value to any such sequence in $S$ by considering it to be an infinite *binary* expansion of a number in the interval $[0, 1]$. For just as any sequence in the standard digits from 0 to 9 can be construed to represent a decimal expansion, so a sequence in the digits 0 and 1 can be construed to represent a binary expansion. For instance, the sequence 12205... produces the decimal expansion value

$$0.12205\ldots = 1 \cdot 10^{-1} + 2 \cdot 10^{-2} + 2 \cdot 10^{-3} + 0 \cdot 10^{-4} + 5 \cdot 10^{-5} + \cdots \in [0, 1],$$

and, completely analogously, the sequence 110101... produces the binary expansion value

$$0.110101\ldots = 1 \cdot 2^{-1} + 1 \cdot 2^{-2} + 0 \cdot 2^{-3} + 1 \cdot 2^{-4} + 0 \cdot 2^{-5} + 1 \cdot 2^{-6} + \ldots$$

which is contained in $[0, 1]$ as well. In summary, what we do is this: we generate an infinite sequence $s \in S$ of zeros and ones by tossing a coin infinitely many times, and then we assign to any such $s$ the value of its corresponding binary expansion which we may denote by, say, $B(s)$. Furthermore, in the light of this definition of $B(s)$, we may now describe the sets $E_k$ in (1.25) as follows: the first of these, namely $E_1 = [0, 1/2]$, is equal to the set of all values $B(s) \in [0, 1]$ for which the first digit in $s$ is equal to zero. This is so because the smallest such expansion is

$$0.0000\ldots = 0,$$

and the largest is

$$0.0111\ldots = \sum_{k=2}^{\infty} \frac{1}{2^k} = \frac{1}{2^2} \sum_{k=0}^{\infty} \frac{1}{2^k} = \frac{1}{4} \cdot \frac{1}{1 - 1/2} = \frac{1}{2} \quad \text{(by (1.11))}.$$

Similarly, the set $E_2 = [0, 1/4] \cup [1/2, 3/4]$ consists of all values $B(s)$ for which the second digit in $s$ is equal to zero. This is so because

$$0.00000\ldots = 0,$$

$$0.00111\ldots = \sum_{k=3}^{\infty} \frac{1}{2^k} = \frac{1}{2^3} \sum_{k=0}^{\infty} \frac{1}{2^k} = \frac{1}{8} \cdot \frac{1}{1 - 1/2} = \frac{1}{4},$$

$$0.10000\ldots = \frac{1}{2}, \text{ and}$$

$$0.10111\ldots = \frac{1}{2} + \sum_{k=3}^{\infty} \frac{1}{2^k} = \frac{1}{2} + \frac{1}{4} = \frac{3}{4}.$$

Proceeding in this manner, we find that in general $E_k$ is the set of all values $B(s)$ for which the $k$-th digit in $s$ is equal to zero. In other words, as each outcome $s$ of our infinite coin-tossing experiment is assigned a value $B(s) \in [0, 1]$, it becomes apparent that the events $E_k$ are independent not only in the abstract mathematical sense of Definition 1.5.3, but also in the concrete independent-trial sense. For as we come to identify each $E_k$ with the set of all 01-sequences $s \in S$ that display a zero in the $k$-th position, and as we construe all the digits in $s$ to be outcomes of individual coin tosses that are performed entirely independently of each other, we also come to see very distinctly how abstract and concrete independence—at least in this present example—coincide perfectly.

For clarity, we need to add that the identification, just referred to, of the events $E_k$ with subsets of $S$ is a special case of a broader identification of arbitrary events $E \subset [0, 1]$ with events in $S$. In fact, it is by means of this more general identification that events in $S$ are defined to be events in the first place. That is to say, if $E$ is an event in $[0, 1]$, then the preimage of $E$ under $B$ in $S$ is the event in $S$ that corresponds to $E$. Moreover, the preimage of $E$ under $B$ is denoted by $B^{-1}(E)$ and is defined to be the set of all $s \in S$ for which $B(s) \in E$. So

$$B^{-1}(E) := \{s \in S \mid B(s) \in E\},$$

is the event in $S$ that corresponds to $E$, and all events in $S$—by definition—are in essence of this form. Furthermore, the probability of $B^{-1}(E)$ is naturally defined to be the probability of $E$, that is,

$$P(B^{-1}(E)) := P(E). \tag{1.26}$$

Given this definition, it is important to understand that our present usage of the symbol $B^{-1}$ for the purpose of denoting preimages is not to be confused with the inverse function notation that is familiar from elementary calculus. In fact, the function $B$ is not one-to-one and therefore not invertible because, for example, $X(1000...) = 1/2 = X(0111...)$. On the surface, this non-invertibility may seem to be problematic. For, if the preimage under $B$ of an event $E \subset [0, 1]$ can contain, so to speak, more elements than $E$ itself, then how are we justified in setting the probability of the former equal to the probability of the latter, as we did in (1.26)? In order to see why this problem is really a non-problem, it is helpful to note to begin with that a number $x \in [0, 1]$ has a non-unique binary expansion only if it has an expansion that is eventually constant in the sense that from some point onwards the digits in its expansion are either all zero or all one. For instance, one expansion of $x = 3/16$ is

$$0.001100000...$$

and another is

$$0.001011111...$$

Moreover, the numbers $x = B(s) \in [0, 1]$ for which $s$ is eventually constant are easily seen to be the binary rational numbers, that is, the numbers $x = k/2^n$ with $0 \le k \le 2^n$. But since the sets $R_n := \{k/2^n \mid 0 \le k \le 2^n\}$ can evidently be listed (the list index is $n$) and since each $R_n$ is finite, it follows that the set $R = \bigcup_{n=1}^{\infty} R_n$ of all binary rationals can be listed as well. By implication, there exists an infinite sequence of values $x_1, x_2, \cdots \in [0, 1]$ such that $\{x_i \mid i \in \mathbb{N}\}$ contains all points in $[0, 1]$ whose binary expansion is not unique. Furthermore, the fact that $P(\{x\}) = 0$ for all $x \in [0, 1]$ allows us to infer that

$$P(\{x_i \mid i \in \mathbb{N}\}) = \sum_{i=1}^{\infty} P(\{x_i\}) = 0,$$

and this in turn shows that the violation by $B$ of the one-to-one mapping property—by being confined to a set of probability zero—is probabilistically irrelevant. That is to say, the defining equation (1.26) is indeed adequate and non-problematic.

# Exercises

**1.5.6.** Assume that $E$ and $F$ are events in a sample space $S$ such that $P(E) = 0.7$, $P(F) = 0.6$ and $P(E \cup F) = 0.8$. Find $P(E|F)$ and $P(F|E)$. Are $E$ and $F$ independent?

**1.5.7.** Assume that $E$ and $F$ are events in a sample space $S$ such that $P(E) = 0.5$, $P(F) = 0.6$ and $P(E \cup F) = 0.8$. Find $P(E|F)$ and $P(F|E)$. Are $E$ and $F$ independent?

**1.5.8.** Assume that $P(E) = P(E|F) = 0.5$ and that $P(F) = 0.4$. Find $P(E \cup F)$.

**1.5.9.** Assume that two events $E$ and $F$ satisfy the following conditions: $P(E), P(F) > 0$ and $P(E \cap F) = 0$. Show that $E$ and $F$ are not independent.

**1.5.10.** An event $E$ is said to be impossible if $P(E) = 0$. Show that an impossible event is independent of any other event.

**1.5.11.** Explain how the concept of independence is relevant to Exercise 1.3.36.

**1.5.12.** If it were the case that 99% of all computer failures are software related and that 3% are hardware related, then what percentage of computer failures would be due to software and hardware problems simultaneously, given that hardware and software failures are not causally linked to each other?

**1.5.13.** Find an example of a sample space $S$ and events $E, F, G \subset S$ such that $E$ and $F$ are independent, $E$ and $G$ are independent, $F$ and $G$ are independent, but $E$, $F$, and $G$ are not independent.

**1.5.14.** A number $k$ is chosen at random from the set $\{1, \ldots, 10\}$ and a second number $i$ is chosen at random from the set $\{k, \ldots, 10\}$. Find the conditional probability $P(E|F)$ for $E := \{(k, i) \mid i + k = 10\}$ and $F := \{(k, i) \mid i - k = 2\}$ and decide whether $E$ and $F$ are independent.

**1.5.15.** Assume that $E$ and $F$ are independent events in a sample space $S$. Show that $S \setminus E$ and $F$ are independent and that $S \setminus E$ and $S \setminus F$ are independent as well.

**1.5.16.** Let $E$ be an event in a sample space $S$. Is it possible for $E$ and $S \setminus E$ to be independent? Explain your answer.

# Chapter 2

# Random Variables and Distributions

## 2.1   Random Variables

In several examples in Chapter 1 we assigned values to random outcomes in a sample space. The number of heads assigned to a finite sequence of coin tosses was one example, the binary expansion value assigned to an infinite sequence of zeros and ones was another, and the distance $d = |\sin(\theta)|$ assigned to a random angle $\theta \in [0, 2\pi]$ was yet another. As it turns out, this kind of assignment of values to outcomes is very frequently encountered and is, in fact, foundational to the entire theory of probability. Consequently, the study of functions of the form $X : S \to \mathbb{R}$ will be from here on onwards a very prominent theme. Admittedly, though, our treatment of these functions will not and cannot be completely rigorous because our prior treatment of events and probability measures was not entirely rigorous either. So we will need to be content to assume that the functions $X$ that we will be working with in standard examples are somehow 'suitable' in that they are compatible with the probability measures defined on their domains.

**2.1.1 Definition.** If $S$ is the sample space of a probability experiment, then any (suitable) function $X : S \to \mathbb{R}$ is said to be a *random variable*.

**2.1.2 Example.** If a coin is tossed twice, so that $S = \{HH, HT, TH, TT\}$, then the equation

$$X(s) := \# \text{ of occurrences of the letter } H \text{ in } s \qquad (2.1)$$

defines a random variable on $S$ where $s$, of course, is an arbitrary outcome in $S$. More precisely, the values of $X$ are given as follows:

$$X(HH) = 2, \ X(HT) = 1, \ X(TH) = 1, \ \text{and, } X(TT) = 0.$$

As a matter of course, this example can be generalized to the case of $n$ successive coin tosses via the same defining equation (2.1). That is to say, if $s$ is a sequence of length $n$ in the letters $H$ and $T$, then $X(s)$ is the number of times that $H$ appears in $s$.

**2.1.3 Example.** If two numbers $s$ and $t$ are chosen at random from the interval $[1,3]$, so that $S = [1,3] \times [1,3]$, then the equation

$$X(s,t) := s^2 t + 2s$$

defines a random variable on $S$. This is not to say that the random variable thus defined is of any genuine interest to us, but it is to say that the notion of a random variable is very broadly inclusive indeed. A random variable is nothing mysterious, and it isn't even inherently 'random'—as its name misleadingly suggests—because it's really just a function defined on a sample space—nothing more and nothing less.

**2.1.4 Definition.** A random variable $X : S \to \mathbb{R}$ is said to be *discrete* if its range consists of a finite or infinite list of discrete values $x_1, x_2, x_3, \ldots$, that is, if $R(X) = \{x_1, x_2, x_3, \ldots\}$.

**2.1.5 Example.** The random variable defined in Example 2.1.2 is discrete because its range is the finite set $\{0, 1, 2\}$, but the random variable in Example 2.1.3 is not discrete because its range happens to be the continuous interval $[3, 33]$.

*Remark.* The alert reader may have noticed that the second answer given in the preceding example is actually not entirely obvious, to say the least. For while it is immediately apparent that the elements of a finite set like $\{0, 1, 2\}$ can indeed be listed, it isn't really clear at all that the same isn't true of the elements in an interval like $[3, 33]$ or, say, $[0, 1]$ if the list is allowed to be infinite in length. After all, there is no end to an infinite listing, by definition, and thus it should be possible to list the elements of any set whatever, be it an interval in $\mathbb{R}$, a square in $\mathbb{R}^2$ or a cube in $\mathbb{R}^3$. It is tempting to think so, but as it turns out, it cannot be done. For if there did exist an infinite listing of values $x_1, x_2, x_3, \ldots$ that exhausts the entire interval from zero to one, that is, $\{x_1, x_2, x_3, \ldots\} = [0, 1]$, then, in representing each $x_k$ by an infinite decimal expansion, we would be able to write down a list of decimal expansions like

$$x_1 = 0.5134\ldots$$
$$x_2 = 0.3431\ldots$$
$$x_3 = 0.1200\ldots$$
$$x_4 = 0.9776\ldots$$

$$\vdots$$

that contains every such expansion. But this is impossible because if we extract from this list the diagonal expansion by taking the first digit from the first expansion, the second digit from the second expansion, the third digit from the third expansion and so forth and then change this resulting expansion in every digit, we obtain an expansion that the list does not contain. Here is how it works: in the list above the diagonal expansion is

$$
\left.\begin{array}{l}
0.\underline{5}134\ldots \\
0.3\underline{4}31\ldots \\
0.12\underline{0}0\ldots \\
0.977\underline{6}\ldots \\
\vdots
\end{array}\right\} \to 0.5406\ldots
$$

and as we change this expansion in some way in every digit to, say, $0.7231\ldots$, we obtain an expansion that differs from $x_1$ in the first digit, from $x_2$ in the second digit, from $x_3$ in the third digit, and so forth. Consequently, the expansion $0.7231\ldots$ is different from all the numbers listed, and therefore, the list is not complete—it doesn't exhaust the entire interval from zero to one. So an interval like $[0, 1]$ is indeed not a discrete set in the sense of Definition 2.1.4 because the order of infinity of its elements is actually larger than the order of infinity of an infinite one-by-one listing. In other words, *infinity can be larger than infinity*, and elements of sets can therefore be impossible to list.

**2.1.6 Example.** If $S = [0, 1]$ and $X(s) := 0$, then $R(X) = \{0\}$ and $X$ is therefore discrete. This very simple example highlights the fact that discreteness is not a property of the domain $S$ of a random variable—the sample space—but rather of its range. In other words, a random variable can be discrete even though the sample space on which it is defined is not. Conversely, however, discreteness of the sample space always implies discreteness of any random variable defined on it, because if $S = \{s_1, s_2, s_3, \ldots\}$, then $R(X) = \{X(s_1), X(s_2), X(s_3), \ldots\}$.

**2.1.7 Example.** Assume that a coin is tossed until for the first time a head is recorded. Then the sample space $S = \{H, TH, TTH, TTTH, \ldots\}$ is discrete, and therefore, the random variable $X : S \to \mathbb{R}$ that is defined by the equation

$$
X(s) := \# \text{ of letters in } s
$$

is discrete as well. As a matter of course, this also follows from the fact that $R(X)$ in this example equals $\mathbb{N}$.

# Exercises

**2.1.8.** Which of the following quantities or conceptions are commonly believed to be continuous?

- time
- radiation energy
- space

**2.1.9.** Which of the following random variables are discrete?

   **a)** $X : \{1, 2, 3\} \to \mathbb{R}$, $X(s) := s$

   **b)** $X : S \to \mathbb{R}$, $X(s) := 1$.

   **c)** $X : [0, 1] \to \mathbb{R}$, $X(s) := s$.

## 2.2 Discrete Distributions

**2.2.1 Definition.** If $X : S \to \mathbb{R}$ is a discrete random variable, and if we denote by $E_x$ the event consisting of all $s \in S$ for which $X(s) = x$, that is,

$$E_x = \{s \in S \mid X(s) = x\},$$

then the (discrete) *density function* or *distribution function* $f : \mathbb{R} \to \mathbb{R}$ of $X$ is defined by the equation

$$f(x) := P(X = x) := P(E_x).$$

Note: this definition, implies that $f(x) = 0$ whenever $x \notin R(X)$ because if $x \notin R(X)$, then $E_x$ is empty.

**2.2.2 Definition.** If $X : S \to \mathbb{R}$ is a discrete random variable, and if we denote by $F_x$ the event consisting of all $s \in S$ for which $X(s) \leq x$, that is,

$$F_x = \{s \in S \mid X(s) \leq x\},$$

then the *cumulative density function* of $X$ is defined by the equation

$$F(x) := P(X \leq x) := P(F_x).$$

Before we look at some examples, we wish to point out that the cumulative density is related to the regular density via the equation

$$F(x) = \sum_{\{y \in R(X) | y \leq x\}} f(y) \tag{2.2}$$

and that the axiomatic equation $P(S) = 1$ therefore implies that

$$1 = \sum_{y \in R(X)} f(y) = \lim_{x \to \infty} F(x)$$

for any discrete random variable $X$.

**2.2.3 Example.** Assume that a biased coin with $P(H) = p$ and $P(T) = 1 - p$ is tossed $n$ times. Then the sample space $S$ consists of all sequences $s$ of length $n$ in the letters $H$ and $T$, and the equation

$$X(s) := \# \text{ of occurrences of the letter } H \text{ in } s$$

defines a discrete random variable on $S$. Using equation (1.7), it is easy to see that the density and cumulative density functions of $X$ satisfy the following equations:

$$f(k) = \binom{n}{k} p^k (1-p)^{n-k}$$

and

$$F(k) = \sum_{i=0}^{k} \binom{n}{i} p^i (1-p)^{n-i}$$

for all $k \in \{0, 1, \ldots, n\} = R(X)$. Furthermore, the density $f$ is referred to as the *binomial density* or *binomial distribution function*.

**2.2.4 Example.** Assume that a biased coin with $P(H) = p$ and $P(T) = 1 - p$ is tossed repeatedly until it shows a head for the first time. Then the density and cumulative density of the random variable $X$, defined in Example 2.1.7, satisfy the following equations:

$$f(k) = (1-p)^{k-1} p$$

and, according to (1.10),

$$F(k) = \sum_{i=1}^{k} (1-p)^{i-1} p = 1 - (1-p)^k$$

for all $k \in \mathbb{N} = R(X)$. Furthermore and by convention, the function $f$ above is referred to as the *geometric density* or *geometric distribution function*.

## Exercises

**2.2.5.** Assume that a number $k$ is selected at random from the set $\{1, \ldots, n\}$ and that a second number $i$ is selected from the set $\{1, \ldots, k\}$.

   a) What is the sample space of this experiment?
   b) What is the density of the random variable $X(k, i) := i$?
   c) What is the density of $Y(k, i) := k$?
   d) What is the density of $Z(k, i) := k - i$?

**2.2.6.** Let $X$ denote the number of times that a die must be rolled until it shows for the first time a 4 or a 5.

**a)** What is the range of $X$?

**b)** Find the density $f$ of $X$.

**c)** Find the cumulative density $F(k)$ of $X$ for all integers $k \in \mathbb{Z}$.

**2.2.7.** Assume that $X$ is the output of a random number generator such that $P(X = k) = 1/10$ for all $k \in \{0, 1, \ldots, 9\}$. Find $F(k)$ for all integers $k \in \mathbb{Z}$.

**2.2.8.** Assume that $f(k) = C/3^k$ is the density of a random variable $X$, the output values of which are $k = 1, 2, 3$.

**a)** Find the value of $C$.

**b)** Find the cumulative density $F(k)$ for all integers $k \in \mathbb{Z}$.

**2.2.9.** Let $s$ be a random value in $[0, 1]$ and let

$$X(s) := \begin{cases} 1 & \text{if } s \in [0, 1/3], \\ 4 & \text{if } s \in (1/3, 5/6], \\ 2 & \text{if } s \in (5/6, 1]. \end{cases}$$

Find $f(x)$ and $F(x)$ for all $x \in \mathbb{R}$.

## 2.3 Expectation and Variance (Discrete Case)

In order to introduce the vital notion of the *expected value* of a random variable, it is helpful to begin with a simple example. So let us assume that we roll a die repeatedly and that we write down numbers according to the following rules: we write down...

- a 1 whenever the die shows a 1 or a 2,
- a 4 whenever the die shows a 3, a 4, or a 5, and
- a 3 whenever the die shows a 6.

In essence, these rules define a random variable $X$ on the sample space $S = \{1, 2, 3, 4, 5, 6\}$ that assigns values to the outcomes in $S$ as follows:

$$X(1) := X(2) := 1, \ X(3) := X(4) := X(5) := 4, \text{ and } X(6) := 3.$$

Given this setup, the defining equation of the density function $f$ of $X$ is

$$f(x) = \begin{cases} P(X = 1) = P(\{1, 2\}) = 1/3 & \text{if } x = 1, \\ P(X = 4) = P(\{3, 4, 5\}) = 1/2 & \text{if } x = 4, \\ P(X = 3) = P(\{6\}) = 1/6 & \text{if } x = 3, \\ 0 & \text{if } x \in \mathbb{R} \smallsetminus \{1, 3, 4\}. \end{cases}$$

That said, the question that we wish to answer is this: what value do we expect to find if we roll a die, say, 600 times, write down numbers according to the rules stated above after each roll, and then take the average of the numbers we wrote down? As it turns out, the answer isn't difficult to find. For given that the probabilities for writing down a 1, a 4, and a 3 are 1/3, 1/2, and 1/6, respectively, we surely expect that ideally 200 out of 600 trials produce a 1, 300 out of 600 a 4, and 100 out of 600 a 3. Thus the average after 600 trials may be expected to be

$$
\begin{aligned}
\frac{200 \cdot 1 + 300 \cdot 4 + 100 \cdot 3}{600} &= 1 \cdot \frac{200}{600} + 4 \cdot \frac{300}{600} + 3 \cdot \frac{200}{600} \\
&= 1 \cdot \frac{1}{3} + 4 \cdot \frac{1}{2} + 3 \cdot \frac{1}{6} \\
&= 1 \cdot P(X = 1) + 4 \cdot P(X = 4) + 3 \cdot P(X = 3) \\
&= 1 f(1) + 4 f(4) + 3 f(3) \\
&= \sum_{x \in R(X)} x f(x).
\end{aligned}
$$

Motivated by this result, we introduce the following definition:

**2.3.1 Definition.** If $X : S \to \mathbb{R}$ is a discrete random variable, then the *expected value* or *expectation* of $X$ is

$$
E(X) := \sum_{x \in R(X)} x f(x).
$$

To obtain a useful modification of this formula for $E(X)$, it is helpful to take another look at our preceding example: if the die is unbiased, as we assumed, then each of the numbers from 1 to 6 in the sample space occurs with probability 1/6. Consequently, we may expect the die to show each of these numbers with the same frequency of 100 out of 600. By implication, we may expect to write down each of the output values from $X(1)$ to $X(6)$ one hundred times, and the average of these output values after 600 trials may therefore be expected to be

$$
\frac{100 X(1) + \cdots + 100 X(6)}{600} = \sum_{k=1}^{6} X(k) \frac{1}{6} = \sum_{k \in S} X(k) P(\{k\}).
$$

This shows that—in the case where the sample space $S$ is discrete—the expected value of a random variable $X : S \to \mathbb{R}$ can be computed from the following alternative formula:

$$
\boxed{E(X) = \sum_{s \in S} X(s) P(\{s\}).}
\tag{2.3}
$$

Note: the expected value $E(X)$ is also referred to as the *mean* or *mean value* of $X$ and is therefore frequently denoted by the Greek letter $\mu$. So we may write

$$\mu = \mu(X) = E(X).$$

**2.3.2 Example.** For the density $f$ defined in Example 2.2.4 and the corresponding random variable $X$ in Example 2.1.7 we find that

$$E(X) = \sum_{k=1}^{\infty} k(1-p)^{k-1}p = p \sum_{k=1}^{\infty} kx^{k-1}\bigg|_{x=1-p} = p \frac{d}{dx} \sum_{k=0}^{\infty} x^k\bigg|_{x=1-p}$$

$$= p \frac{d}{dx} \frac{1}{1-x}\bigg|_{x=1-p} = \frac{1}{p}.$$

**2.3.3 Example.** If a biased coin with $P(H) = p$ and $P(T) = 1 - p$ is tossed $n$ times, we naturally expect that the number of heads is $np$. In other words, $np$ ought to be the expected value of the random variable $X$ in Example 2.2.3. And indeed, using the formula in Definition 2.3.1, we find that

$$E(X) = \sum_{k=0}^{n} k\binom{n}{k} p^k (1-p)^{n-k} = \sum_{k=1}^{n} \frac{n!}{(k-1)!(n-k)!} p^k (1-p)^{n-k}$$

$$= np \sum_{k=1}^{n} \binom{n-1}{k-1} p^{k-1}(1-p)^{n-1-(k-1)}$$

$$= np \sum_{k=0}^{n-1} \binom{n-1}{k} p^k (1-p)^{n-1-k} = np$$

because

$$\sum_{k=0}^{n-1} \binom{n-1}{k} p^k (1-p)^{n-1-k}$$

is the total probability of a binomial distribution with parameters $n-1$ and $p$ and is therefore equal to one.

**2.3.4 Theorem.** *If $X : S \to \mathbb{R}$ is constant, that is, $X(s) = c$ for all $s \in S$, then $E(X) = c$.*

*Proof.* If $X(s) = c$ for all $s \in S$, then $R(X) = \{c\}$ and the density $f$ of $X$ satisfies the equation $f(c) = P(X = c) = 1$. Hence

$$E(X) = \sum_{x \in R(X)} xf(x) = cf(c) = c,$$

as desired. □

**2.3.5 Theorem.** *If $X, Y : S \to \mathbb{R}$ are (discrete) random variables and $\lambda \in \mathbb{R}$, then*

a) $E(X + Y) = E(X) + E(Y)$ *and*

b) $E(\lambda X) = \lambda E(X)$.

*Proof.* **a)** This property holds in general, but for convenience we will assume that $S$ is discrete:

$$E(X + Y) = \sum_{s \in S}(X(s) + Y(s))P(s) = \sum_{s \in S} X(s)P(s) + \sum_{s \in S} Y(s)P(s)$$
$$= E(X) + E(Y).$$

**b)** If $\lambda = 0$, then $\lambda X = 0$ and $E(\lambda X) = 0 = \lambda E(X)$ by Theorem 2.3.4. If $\lambda \neq 0$, then $R(\lambda X) = \{\lambda x \mid x \in R(X)\}$ and for all $\lambda x \in R(\lambda X)$ we have $P(\lambda X = \lambda x) = P(X = x) = f(x)$. Hence

$$E(\lambda X) = \sum_{x \in R(X)} \lambda x f(x) = \lambda \sum_{x \in R(X)} x f(x) = \lambda E(X)$$

as desired.                                                                       □

In working with random experiments, it is helpful not only to know the expected values of any pertinent random variables but also the expected deviations from these values. However, exactly what we mean by 'deviation' is not inherently obvious and different modes of interpretation can plausibly be entertained. The most obvious meaning that we can give to the word 'deviation', in a somewhat more narrow, mathematical sense, is that of a numerical difference. So the expected deviation of $X$ from its expected value $E(X)$ could simply be construed to be the expected value of $X - E(X)$. However, according to Theorems 2.3.4 and 2.3.5 (with $c = E(X)$), it is the case that

$$E(X - E(X)) = E(X + (-1)E(X)) = E(X) + E((-1)E(X))$$
$$= E(X) + (-1)E(E(X)) = E(X) - E(E(X))$$
$$= E(X) - E(X) = 0.$$

So this is not meaningful because a deviation that is always equal to zero provides no information at all. A more promising alternative is to consider the expected value of the absolute value of the difference $X - E(X)$. Unfortunately, though, absolute values are algebraically cumbersome to work with, and the absolute value function $g(x) = |x|$ is also not differentiable at zero. For this reason, mathematicians have decided to use the square rather than the absolute value of the difference. Hence we introduce the following definition:

**2.3.6 Definition.** If $X : S \to \mathbb{R}$ is a random variable, then

$$\text{Var}(X) := E((X - E(X))^2)$$

is said to be the *variance* of $X$ and is also commonly denoted by $\sigma^2(X)$ or simply $\sigma^2$. Furthermore, the value $\sigma = \sigma(X) = \sqrt{\text{Var}(X)}$ is referred to as the *standard deviation* of $X$.

**2.3.7 Theorem.** *For any random variable $X : S \to \mathbb{R}$ and all scalars $\lambda, c \in \mathbb{R}$ it is the case that*

a) $\text{Var}(X) = E(X^2) - E(X)^2$,

b) $\text{Var}(\lambda X) = \lambda^2 \text{Var}(X)$, *and*

c) $\text{Var}(X + c) = \text{Var}(X)$.

*Proof.* **a)**

$$\text{Var}(X) = E((X - E(X))^2) = E(X^2 - 2XE(X) + E(X)^2)$$
$$= E(X^2) - 2E(X)E(X) + E(X)^2 = E(X^2) - E(X)^2.$$

**b)** This is a trivial consequence of a) and Theorem 2.3.5b.
**c)** $\text{Var}(X + c) = E((X + c - E(X + c))^2) = E((X - E(X))^2) = \text{Var}(X)$ by Theorems 2.3.4 and 2.3.5. □

Given the statement of the preceding theorem, it may be helpful to point out that

$$E(X^2) = \sum_{x \in R(X)} x^2 f(x)$$

or, more generally, that

$$\boxed{E(g(X)) = \sum_{x \in R(X)} g(x)f(x)} \tag{2.4}$$

for any function $g : \mathbb{R} \to \mathbb{R}$ (for which the sum $\sum_{x \in R(X)} g(x)f(x)$ is convergent in the case where $R(X)$ is infinite). To see why this is so, we need to understand to begin with that the composition $g \circ X = g(X)$ is a function from the sample space $S$ on which $X$ is defined into the range of $g$. Since the range of $g$ is contained in $\mathbb{R}$, it follows that the composition of $g$ and $X$ is a function from $S$ into $\mathbb{R}$ and thus itself a random variable. Furthermore, if $X$ is discrete, then so is $g \circ X = g(X)$ because if $R(X) = \{x_1, x_2, \ldots\}$, then

$$R(g(X)) = \{g(x_1), g(x_2), \ldots\}. \tag{2.5}$$

While this may seem completely obvious, we need to be careful to understand that $g$ may assign the same output value to different input values and that therefore some of the output values in the list $g(x_1), g(x_2), \ldots$ may be equal. Fortunately, though, this possibility does not compromise the validity of (2.5) because if elements in a set are listed more than once, the set remains unchanged (for instance, $\{1, 1, 2, 3, 3\} = \{1, 2, 3\}$). However, the fact just mentioned that $g$ can possibly assign the same output value to different input values makes the seemingly simple and self-evident equation (2.4) a bit more difficult to verify than one might naively expect. To be sure, the formula in Definition 2.3.1 readily implies that

$$E(g(X)) = \sum_{y \in R(g(X))} yP(g(X) = y),$$

but showing that the sum on the right of this equation is equal to the sum in (2.4) takes a bit of additional explanation. The way to proceed here is to observe that for any $y \in R(g(X))$ the event $F_y = \{s \in S \mid g(X(s)) = y\}$ is a disjoint union of all the events $E_k := \{s \in S \mid X(s) = x_k\}$ for which $g(x_k) = y$. For given this observation, it follows that

$$P(g(X) = y) = P(F_y) = \sum_{\{k \in \mathbb{N} \mid g(x_k) = y\}} P(E_k) = \sum_{\{k \in \mathbb{N} \mid g(x_k) = y\}} f(x_k),$$

and therefore,

$$E(g(X)) = \sum_{y \in R(g(X))} yP(g(X) = y) = \sum_{y \in R(g(X))} \sum_{\{k \in \mathbb{N} \mid g(x_k) = y\}} yf(x_k)$$

$$= \sum_{y \in R(g(X))} \sum_{\{k \in \mathbb{N} \mid g(x_k) = y\}} g(x_k)f(x_k) = \sum_{x \in R(X)} g(x)f(x),$$

as desired.

**2.3.8 Example.** If $X : \{1, 2, 3, 4, 5, 6\} \to \{1, 3, 4\}$ is the random variable defined on p.47, then $f(1) = P(X = 1) = 1/3$, $f(3) = P(X = 3) = 1/6$, and $f(4) = P(X = 4) = 1/2$. Hence

$$E(X) = 1 \cdot f(1) + 3 \cdot f(3) + 4 \cdot f(4) = \frac{1}{3} + \frac{3}{6} + \frac{4}{2} = \frac{17}{6},$$

$$E(X^2) = 1^2 \cdot f(1) + 3^2 \cdot f(3) + 4^2 \cdot f(4) = \frac{1}{3} + \frac{9}{6} + \frac{16}{2} = \frac{59}{6},$$

and

$$\text{Var}(X) = E(X^2) - E(X)^2 = \frac{59}{6} - \left(\frac{17}{6}\right)^2 = \frac{65}{36}.$$

**2.3.9 Example.** The variance of the geometric random variable $X$ in Example 2.3.2 is

$$\text{Var}(X) = E(X^2) - E(X)^2 = \sum_{k=1}^{\infty} k^2(1-p)^{k-1}p - \frac{1}{p^2}.$$

In order to evaluate the series on the right, we observe that

$$\frac{x}{(1-x)^2} = x\frac{d}{dx}\frac{1}{1-x} = x\sum_{k=1}^{\infty} kx^{k-1} = \sum_{k=1}^{\infty} kx^k$$

and that, by implication,

$$\frac{1+x}{(1-x)^3} = \frac{d}{dx}\frac{x}{(1-x)^2} = \sum_{k=1}^{\infty} k^2 x^{k-1}$$

for all $x \in (-1, 1)$. Hence

$$\text{Var}(X) = \frac{p(1+(1-p))}{(1-(1-p))^3} - \frac{1}{p^2} = \frac{2-p}{p^2} - \frac{1}{p^2} = \frac{1-p}{p^2}.$$

**2.3.10 Example.** The variance of the binomial random variable in Example 2.3.3 is

$$\text{Var}(X) = E(X^2) - E(X)^2 = \sum_{k=1}^{n} k^2 \binom{n}{k} p^k (1-p)^{n-k} - n^2 p^2$$

$$= \sum_{k=1}^{n} k(k-1)\binom{n}{k} p^k (1-p)^{n-k} + \sum_{k=1}^{n} k\binom{n}{k} p^k (1-p)^{n-k} - n^2 p^2$$

$$= \sum_{k=2}^{n} k(k-1)\binom{n}{k} p^k (1-p)^{n-k} + E(X) - n^2 p^2$$

$$= \sum_{k=2}^{n} \frac{n!}{(k-2)!(n-k)!} p^k (1-p)^{n-k} + np - n^2 p^2$$

$$= n(n-1)p^2 \sum_{k=2}^{n} \frac{(n-2)!}{(k-2)!((n-2)-(k-2))!} p^{k-2} (1-p)^{(n-2)-(k-2)}$$
$$+ np - n^2 p^2$$

$$= n(n-1)p^2 \sum_{k=0}^{n-2} \frac{(n-2)!}{k!((n-2)-k)!} p^k (1-p)^{(n-2)-k} + np - n^2 p^2$$

$$= n(n-1)p^2 \sum_{k=0}^{n-2} \binom{n-2}{k} p^k (1-p)^{(n-2)-k} + np - n^2 p^2$$

$$= n(n-1)p^2 + np - n^2 p^2 = np(1-p).$$

The preceding example illustrates quite strikingly that calculations of expected values of powers of $X$ (like $X^2$) by means of the formula in Definition 2.3.1 can easily become fairly complex and cumbersome. For this reason we will now introduce the so-called *moment-generating* function as a computational tool that often makes these calculations more convenient and efficient.

**2.3.11 Definition.** If $X : S \to \mathbb{R}$ is a (discrete) random variable and $f$ is the density of $X$, then

$$m_X(t) := E\left(e^{tX}\right) = \sum_{x \in R(X)} e^{tx} f(x)$$

is said to be the moment-generating function of $X$.

**2.3.12 Theorem.** *For all positive integers $k$ we have*

$$\left.\frac{d^k}{dt^k} m_X(t)\right|_{t=0} = E(X^k).$$

*Proof.* Let $f$ be the (discrete) density of $X$. Then, for $k = 1$, we find that

$$\frac{d}{dt} m_X(t) = \frac{d}{dt} E\left(e^{tX}\right) = \frac{d}{dt} \sum_{x \in R(X)} e^{tx} f(x) = \sum_{x \in R(X)} x e^{tx} f(x),$$

and therefore,

$$\left.\frac{d}{dt} m_X(t)\right|_{t=0} = \sum_{x \in R(X)} x f(x) = E(X).$$

Taking another derivative and setting $k$ equal to 2, it follows that

$$\frac{d^2}{dt^2} m_X(t) = \frac{d}{dt}\left(\frac{d}{dt} m_X(t)\right) = \frac{d}{dt} \sum_{x \in R(X)} x e^{tx} f(x) = \sum_{x \in R(X)} x^2 e^{tx} f(x),$$

and therefore,

$$\left.\frac{d^2}{dt^2} m_X(t)\right|_{t=0} = \sum_{x \in R(X)} x^2 f(x) = E(X^2).$$

Proceeding in this fashion, it is not difficult to see that in general

$$\left.\frac{d^k}{dt^k} m_X(t)\right|_{t=0} = E(X^k)$$

for all $k \in \mathbb{N}$. $\qquad\qquad\qquad\qquad\qquad\qquad\qquad\qquad\qquad\qquad\square$

**2.3.13 Example.** Considering again the geometric random variable $X$ in Example 2.3.9, we find that

$$m_X(t) = E(e^{tX}) = \sum_{k=1}^{\infty} e^{tk}(1-p)^{k-1} p = p e^t \sum_{k=0}^{\infty} (e^t(1-p))^k = \frac{p e^t}{1 - e^t(1-p)}.$$

Hence

$$m'_X(t) = \frac{pe^t(1 - e^t(1-p)) + p(1-p)e^{2t}}{(1 - e^t(1-p))^2} = \frac{pe^t}{(1 - e^t(1-p))^2}$$

and

$$m''_X(t) = \frac{pe^t(1 - e^t(1-p))^2 + 2p(1-p)e^{2t}(1 - e^t(1-p))}{(1 - e^t(1-p))^4}$$
$$= \frac{pe^t(1 + (1-p)e^t)}{(1 - e^t(1-p))^3}.$$

Using Theorem 2.3.12, it follows that

$$E(X) = m'_X(0) = \frac{p}{(1 - (1-p))^2} = \frac{1}{p},$$

and

$$\mathrm{Var}(X) = E(X^2) - E(X)^2 = m''_X(0) - \frac{1}{p^2} = \frac{p(1 + (1-p))}{(1 - (1-p))^3} - \frac{1}{p^2} = \frac{1-p}{p^2},$$

as desired (see Example 2.3.9).

**2.3.14 Example.** Let $X$ be the binomial random variable that we considered in Example 2.3.10. Then the binomial formula (1.5) (with $a = e^t p$ and $b = 1 - p$) implies that

$$m_X(t) = \sum_{k=0}^{n} e^{tk} \binom{n}{k} p^k (1-p)^{n-k} = \sum_{k=0}^{n} \binom{n}{k} (e^t p)^k (1-p)^{n-k}$$
$$= (e^t p + 1 - p)^n.$$

Hence

$$m'_X(t) = npe^t(e^t p + 1 - p)^{n-1},$$

and

$$m''_X(t) = npe^t(e^t p + 1 - p)^{n-1} + n(n-1)p^2 e^{2t}(e^t p + 1 - p)^{n-2}.$$

Consequently, Theorem 2.3.12 allows us to infer that

$$E(X) = m'_X(0) = np$$

and

$$\mathrm{Var}(X) = E(X^2) - E(X)^2 = m''_X(0) - n^2 p^2 = np + n(n-1)p^2 - n^2 p^2$$
$$= np(1-p),$$

as desired (see Example 2.3.10).

**2.3.15 Example.** Let's take another look at the binomial random variable $X$ in the preceding example, but let's denote it this time by $X_n$ in order to make more apparent the dependence of $X$ on $n$. The sample space $S_n$ of $X_n$ consists of all the sequences of length $n$ in the letters $H$ and $T$. Consequently, if we define

$$X_{n,k}(s) := \begin{cases} 1 & \text{if the } k\text{-th letter in } s \text{ is an } H, \\ 0 & \text{if the } k\text{-th letter in } s \text{ is a } T, \end{cases}$$

for all $s \in S_n$, then

$$X_n = X_{n,1} + \cdots + X_{n,n} = \sum_{k=1}^{n} X_{n,k},$$

and Theorem 2.3.5b thus implies that

$$E\left(\frac{1}{n}\sum_{k=1}^{n} X_{n,k}\right) = E\left(\frac{X_n}{n}\right) = \frac{E(X_n)}{n} = p.$$

Furthermore, according to Theorem 2.3.7b, it is the case that

$$\mathrm{Var}\left(\frac{1}{n}\sum_{k=1}^{n} X_{n,k}\right) = \mathrm{Var}\left(\frac{X_n}{n}\right) = \frac{\mathrm{Var}(X_n)}{n^2} = \frac{np(1-p)}{n^2} = \frac{p(1-p)}{n},$$

and therefore,

$$\lim_{n\to\infty} \mathrm{Var}\left(\frac{1}{n}\sum_{k=1}^{n} X_{n,k}\right) = 0.$$

So the variance—and the standard deviation therewith—of the average of the random variables $X_{n,k}$ shrinks to zero as $n$ tends to infinity. But how is this significant? To answer this question, we need to understand that $p$ is the expected value not only of the averages $(X_{n,1} + \cdots + X_{n,n})/n$ but also of each individual random variable $X_{n,k}$. This fact can be deduced by setting $n = 1$, but it can also be verified directly:

$$E(X_{n,k}) = 0 \cdot P(0) + 1 \cdot P(1) = 1 \cdot P(H) = p.$$

However, exactly why do we say that $p$ is the expected value of $X_{n,k}$ if the only values that $X_{n,k}$ can ever assume are 0 and 1 and if both of these values are different from $p$? What does it mean to say that the expected value of a random variable is a value that the random variable can never produce and that therefore cannot be expected to occur? Well, it means that in order for the notion of an *expected* value to make any sense at all it must somehow be tied to the notion of an average taken over a large number of trials—and the larger the number the better the resulting approximation of the expected value $p$:

when we perform 10 trials and take the average we don't expect to get as good an approximation for the ideal mean $p$ as when we perform 100 or 1000 trials. So that's our intuitive understanding of the nature of this kind of simple random process in which trials are performed *idependently* in an identical physical fashion (using always the same coin or the same die in the same procedural manner). But what is important here to realize is that this natural intuitive understanding of ours is very nicely confirmed by the rigorous mathematical fact of the contraction of the variance. For insofar as the variance is rightly considered to be a measure for the deviation from the mean, its contraction to zero as the number of terms in the average tends to infinity may rightly be viewed to be directly expressive of our pertinent intuitive expectation that the averages are likely to approximate the ideal mean $p$ more and more closely as this number becomes larger and larger. Furthermore and very importantly, the contraction of the variance is a universal mathematical proposition that depends for its validity—as we shall see in Section 2.9—only on the independence and procedural identity of the random trials that produce the outcomes that are averaged. So what we here are dealing with is not an inconsequential assertion about the vanishing of a limit in one particular example, but rather a broadly applicable, fundamental principle that vitally informs the entire theory of probability.

# Exercises

**2.3.16.** Find the expected value of the random variable $X$ defined in Exercise 2.2.5.

**2.3.17.** Use the formula for the geometric series to determine $E(X)$ for the random variable $X$ defined in Exercise 2.2.6.

**2.3.18.** Determine $E(X)$ for the random variable $X$ defined in Exercise 2.2.7.

**2.3.19.** Assume that $f(k) = C/(k(k+1))$ is the density of a random variable $X$, the output values of which are the positive integers.

    **a)** Find the value of $C$.

    **b)** What is the expected value of $X$?

**2.3.20.** Assume that $f(k) = C/(k(k+1)(k+2))$ is the density of a random variable $X$, the output values of which are the positive integers.

    **a)** Find the value of $C$.

    **b)** What is the expected value of $X$?

    **c)** What is the variance of $X$?

**2.3.21.** Assume that $X$ is a discrete random variable with range $R(X) = \mathbb{N}$. If $f(k)$ is the density of $X$, then what are the density functions of $2X$ and $X^2$ in terms of $f$ and what are the corresponding expected values?

**2.3.22.** Which of the following equations is true for all (discrete) random variables $X$? Explain your answer.

a) $E(2X) = 2E(X)$

b) $E(X^2) = E(X)^2$

**2.3.23.** Assume that $f(k) = Ck/3^k$ is the density of a random variable $X$, the output values of which are the positive integers.

a) Find the value of $C$.

b) Find the moment-generating function of $X$.

c) Use b) to determine $E(X)$ and $\mathrm{Var}(X)$.

**2.3.24.** Assume that $P(X = 0) = p^2$, $P(X = 1) = 2p(1 - p)$, and $R(X) = \{0, 1, 2\}$.

a) Find $P(X = 2)$.

b) Find the moment-generating function of $X$.

c) Use b) to find the expectation and variance of $X$.

**2.3.25.** Assume that $X$ is a binomial random variable with parameters $n$ and $p$ such that $E(X) = 8$ and $E(X^2) = 70$. Find $n$ and $p$

**2.3.26.** What is the smallest value $n$ for which there exists a binomial random variable $X$ for which $\mathrm{Var}(X) = 12$.

**2.3.27.** A die is rolled once and shows a number $k \in \{1, \ldots, 6\}$. Subsequently, the die is rolled $k$ more times and the sum $X$ of the $k + 1$ numbers thus produced is recorded. Find the expected value of $X$.

## 2.4  The Poisson Distribution

Let us assume that we are given an infinite (continuous) reservoir $R$ in which particles—animate or inanimate, real or symbolic—are floating with an average density $\delta$. Given this assumption, we wish to determine the probability that a (continuous) sample of size $r$ contains $X = k$ particles. Since the reservoir is infinite in size, the sample space $S = \{0, 1, 2, \ldots\} = \mathbb{N} \cup \{0\}$ is infinite as well. Furthermore, the random variable $X : S \to \mathbb{R}$ that measures the number of particles in a sample of size $r$ is the identity on $S$, that is, $X(k) = k$ for all $k \in S$.

As a matter of course, the assumption that the reservoir is infinite in size is an idealization, but from a practical computational point of view it usually makes little sense to distinguish between a reservoir that is infinite and one that is finite but still very large. For instance, when we count the number of microorganisms in a few drops of lake water or the dust particles in a small volume of air, the surrounding reservoir is so large compared to the sample that we may as well construe it to be infinite.

In order to find the density of $X$, we imagine $R$ to be broken up into disjoint continuous segments of size $r$ (for example, if $R = \mathbb{R}$, then $R$ would be broken up, in familiar calculus fashion, into disjoint intervals of length $r$), and instead of the entire infinite reservoir, we first consider a finite subreservoir that consists of $n$ segments of size $r$ in which $N$ particles are floating with overall density

$$\delta = \frac{N}{nr}.$$

(Note: this equation obviously can only be satisfied if $r\delta$ is a rational number, but from a practical point of view, this assumption is clearly acceptable.) So as we now pick one of these $n$ segments as our sample of size $r$, the question that we need to answer is this: what is the probability that $k$ particles out of a total of $N$ particles are contained in the chosen sample? Clearly, since the subreservoir consists of $n$ segments of equal size $r$, the probability for any given particle to be contained in the sample is $p = 1/n$ and the corresponding complementary probability for it to be in any of the other $n - 1$ segments is $1 - p = 1 - 1/n$. Given this observation, we may use the formula for the binomial distribution in Example 2.2.3, to infer that the probability for the chosen sample to contain $k$ particles is

$$f_n(k) = \binom{N}{k}\left(\frac{1}{n}\right)^k\left(1 - \frac{1}{n}\right)^{N-k} = \binom{r\delta n}{k}\left(\frac{1}{n}\right)^k\left(1 - \frac{1}{n}\right)^{r\delta n - k}.$$

Thus the density of $X$—the density for the entire infinite reservoir—is

$$f(k) = \lim_{n\to\infty} f_n(k) = \lim_{n\to\infty} \binom{r\delta n}{k}\left(\frac{1}{n}\right)^k\left(1 - \frac{1}{n}\right)^{r\delta n - k}.$$

Using the familiar calculus formula

$$e^y = \lim_{x\to\infty}\left(1 + \frac{y}{x}\right)^x, \tag{2.6}$$

(see for instance [B1], p.224) with $y = -1$ and $x = n$, we find that

$$f(k) = e^{-r\delta}\lim_{n\to\infty}\binom{r\delta n}{k}\left(\frac{1}{n}\right)^k\left(1 - \frac{1}{n}\right)^{-k} = e^{-r\delta}\lim_{n\to\infty}\binom{r\delta n}{k}\left(\frac{1}{n-1}\right)^k$$

$$= \frac{e^{-r\delta}}{k!}\lim_{n\to\infty}\frac{r\delta n(r\delta n - 1)\cdots(r\delta n - k + 1)}{(n-1)^k}.$$

Furthermore, since

$$\lim_{n \to \infty} \frac{r\delta n - c}{n - 1} = r\delta \quad \text{(by, for instance, L'Hôpital's rule)}$$

for any constant $c \in \mathbb{R}$, it follows that

$$\boxed{f(k) = \frac{e^{-r\delta}(r\delta)^k}{k!}} \tag{2.7}$$

for all $k \in S = \mathbb{N} \cup \{0\}$. Given this formula for $f$ and given the familiar series representation of the exponential function ($e^x = \sum_{k=0}^{\infty} x^k/k!$ for all $x \in \mathbb{R}$), the corresponding moment-generating function is

$$m_X(t) = \sum_{k=0}^{\infty} \frac{e^{tk}e^{-r\delta}(r\delta)^k}{k!} = e^{-r\delta} \sum_{k=0}^{\infty} \frac{(e^t r\delta)^k}{k!} = e^{-r\delta}e^{e^t r\delta}. \tag{2.8}$$

**2.4.1 Example.** In 2010 the car-accident death rate per 100 million vehicle miles was 1.11 on average. Given this information, we may ask the following natural question: what was the probability for a given driver to get involved in a fatal accident in 2010 (fatal for either the driver him- or herself or one of the passengers) if he or she drove a total of 12000 miles in 2010? To answer this question, we use the Poisson distribution with $r = 12000$, $\delta = 1.11/100,000,000$, and $k = 0$. This yields the value

$$f(0) = \frac{e^{-12000 \cdot 1.11/100,000,000}(r\delta)^0}{0!} \approx 0.99987,$$

and the probability in question therefore is the complementary probability

$$1 - f(0) \approx 0.00013$$

In other words, the chance of getting involved in a fatal accident in 2010 was about 1.3 in 10000.

# Exercises

**2.4.2.** Use (2.8) to find the mean and variance of $X$ and explain how the values that you find are consistent with the corresponding values for the mean and variance of a binomial random variable with parameters $n$ and $p$.

**2.4.3.** In 2010, the death rate per 100 million vehicle miles was 1.11. What was the probability, in 2010, for a person who drives $x$ miles per year to get involved in a fatal accident?

**2.4.4.** The surface layer of a lake contains bacteria at a density of $1.5 \cdot 10^6$ cells per $ml$ (this is a realistic value). What is the probability that a 6 $ml$ sample contains no more than $9 \cdot 10^6 + 3000$ cells and no fewer than $9 \cdot 10^6 - 3000$ cells? What are the expectation and variance of the random variable that describes this sample?

## 2.5 Continuous Distributions

If a computer produces random values between, say, 0 and 4 with 10-digit precision, then, as we previously discussed (p.18), we may naturally construe the sample space $S$ to be the continuous interval $[0, 4]$. Furthermore, if we assign to every random value $s \in [0, 4]$, for simplicity's sake and for example, the value $s + 1$, then each of these assigned values may be considered to be an output value of the random variable $X : [0, 4] \to \mathbb{R}$, $X(s) := s + 1$. Since the range of this random variable is evidently the interval $[1, 5]$, it follows that $X$ is not discrete. Consequently, the density-defining equation

$$f(x) = P(X = x) = P(\{s \in S \mid X(s) = x\})$$

is in this case inapplicable because for any $x \in [1, 5] = R(X)$ we have

$$P(\{s \in S \mid X(s) = x\}) = P(\{s \in [0, 4] \mid s + 1 = x\}) = P(\{x - 1\}) = 0.$$

However, we already know how to handle this problem because we discussed it earlier in respect of continuous sample spaces like intervals in $\mathbb{R}$ or regions in $\mathbb{R}^2$. So we will do now what we did then and replace single points with extended sets. That is to say, instead of asking how likely it is for $X(s)$ to be equal to $x$, we will inquire how probable it is for $X(s)$ to be contained in an extended interval $[a, b]$. Moreover, since the range of $X$ in our present example is $[1, 5]$ and since $X$ is understood to assume all values $x = s + 1 \in [1, 5]$ with equal probability, we readily find the answer to the latter question to be this:

$$P(\{s \in S \mid X(s) \in [a, b]\}) = P(X \in [a, b]) = P(a \le X \le b)$$
$$= \frac{\text{length of } [1, 5] \cap [a, b]}{\text{length of } [1, 5]} = \frac{\text{length of } [1, 5] \cap [a, b]}{4}.$$

This is progress and it does make sense, but it doesn't really tell us yet what the density of $X$ might be as an actual function $f$ defined on $\mathbb{R}$. However, what it does allow us to do is to define a cumulative density $F$ by setting $a$ equal to $-\infty$ and $b$ equal to $x$:

$$F(x) := P(X \le x) = \frac{\text{length of } [1, 5] \cap (-\infty, x]}{4}$$

$$= \begin{cases} 0/4 = 0 & \text{if } x < 1, \\ (x - 1)/4 & \text{if } 1 \le x \le 5, \\ 4/4 = 1 & \text{if } x > 5. \end{cases} \tag{2.9}$$

Furthermore, given the relation between $F$ and $f$ in the discrete case where we found $F(x)$ to be the sum of all the values $f(y)$ for which $y \in R(X)$ and

$y \leq x$ (see (2.2)), it is natural to require in the present continuous case that $F(x)$ be the *integral* of $f$ from $-\infty$ to $x$, that is,

$$F(x) = P(X \leq x) = \int_{-\infty}^{x} f(t)\, dt. \tag{2.10}$$

Using the fundamental theorem of calculus, it follows that

$$F'(x) = f(x), \tag{2.11}$$

and it this equation, in conjunction with the defining equation

$$F(x) := P(X \leq x) = P(\{s \in S \mid X(s) \leq x\}), \tag{2.12}$$

that may be considered to be the defining equation of the density $f$ of a continuous random variable $X : S \rightarrow \mathbb{R}$. To be sure, equation (2.11) is somewhat problematic as a defining equation for $f$ in our present example because the cumulative density $F$, as given in (2.9), is not differentiable at 1 and 5. However, what we may conclude from (2.9) is that

$$f(x) = \begin{cases} 1/4 & \text{if } x \in (1,5), \\ 0 & \text{if } x \in \mathbb{R} \smallsetminus [1,5] \end{cases} \tag{2.13}$$

and the graph of $f$ therefore looks as follows:

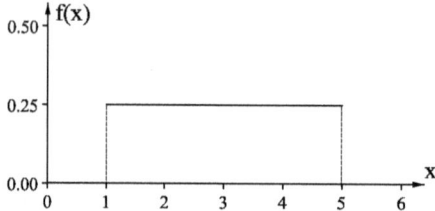

Figure 2.1: graph of the density $f$.

Fortunately, the fact that equation (2.13) does not define $f$ at $x = 1$ and $x = 5$ is not very important, because it never effects any integral involving $f$. For instance, if we ask what the probability might be for $X$ to be contained in an interval $[a, b]$, then equation (2.10) yields the following answer:

$$P(X \in [a,b]) = F(b) - F(a) = \int_{a}^{b} f(x)\, dx. \tag{2.14}$$

So probabilities are integrals, and since integrals of (positive) functions represent areas, itreally doesn't matter whether there are finitely many points at which $f$ is either undefined or given some arbitrary value because the area under a graph over an isolated point is always equal to zero.

The fact that probabilities are integrals and therefore areas under graphs is nicely confirmed by the observation that the total area—or total probability—under the graph of $f$ in Figure 2.1 is equal to $0.25 \cdot (5 - 1) = 1$. To generalize this observation, we now establish the following theorem:

**2.5.1 Theorem.** *If $f$ is the density of a continuous random variable $X : S \to \mathbb{R}$, then $f(x) \geq 0$ for all $x \in \mathbb{R}$ and*

$$\int_{-\infty}^{\infty} f(x)\,dx = 1.$$

*Proof.* Assume that $x$ and $y$ are real numbers such that $x \leq y$. Then

$$\{s \in S \mid X(s) \leq x\} \subset \{s \in S \mid X(s) \leq y\},$$

and therefore, $F(x) = P(X \leq x) \leq P(X \leq y) = F(y)$. This shows that $F$ is increasing and that its derivative $F' = f$, by implication, must be greater than or equal to zero. Moreover, since $S = \{s \in S \mid X(s) \leq \infty\}$, we may apply (2.10) to infer that

$$1 = P(S) = P(X \leq \infty) = \int_{-\infty}^{\infty} f(x)\,dx,$$

as desired. □

To further elucidate the meaning and significance of a density function, we may replace the interval $[a, b]$ in (2.14) with an interval of the form $[x, x + \Delta x]$ and then take the limit as $\Delta x$ converges to zero to obtain the infinitesimal interval $[x, x + dx]$. This yields

$$\frac{P(X \in [x, x + dx])}{dx} = \lim_{\Delta x \to 0^+} \frac{P(X \in [x, x + \Delta x])}{\Delta x} = \lim_{\Delta x \to 0^+} \frac{F(x + \Delta x) - F(x)}{\Delta x}$$
$$= F'(x) = f(x)$$

or, somewhat informally,

$$\boxed{P(X \in [x, x + dx]) = f(x)\,dx.} \tag{2.15}$$

So in the continuous case, it is not the value $f(x)$ itself that represents a probability but rather the product $f(x)\,dx$. For it is this infinitesimal value that equals the probability for $X$ to be contained in the infinitesimal interval $[x, x + dx]$.

Having thus identified the basic defining properties of densities of continuous random variables, we are now ready to address the problem of computing expected values. To do so, we assume that the output of a random experiment is described by a random variable $X : S \to \mathbb{R}$ whose density is $f$. Proceeding in standard calculus fashion, we break up the real axis $\mathbb{R}$ on which $f$ is defined into infinitely many intervals $[x_k, x_{k+1}]$ with $x_k = k\Delta x$ for all positive and negative integers $k$ (i.e., for all $k \in \mathbb{Z}$):

$$\vdots$$

$$[x_{-2}, x_{-1}] = [-2\Delta x, -\Delta x],$$
$$[x_{-1}, x_0] = [-\Delta x, 0],$$
$$[x_0, x_1] = [0, \Delta x],$$
$$[x_1, x_2] = [\Delta x, 2\Delta x],$$
$$[x_2, x_3] = [2\Delta x, 3\Delta x],$$

$$\vdots$$

If $\Delta x$ is small, then, $x \approx x_k$ for all $x \in [x_k, x_{k+1}]$, and the probability $P(X \in [x_k, x_{k+1}])$ may thus be considered to be the continuous case equivalent of the discrete case probability $P(X = x_k)$. By implication, the expected value of $X$ is approximately equal to

$$\sum_{k=-\infty}^{\infty} x_k P(X \in [x_k, x_{k+1}]).$$

Furthermore, using (2.14), we find that

$$P(X \in [x_k, x_{k+1}]) = \int_{x_k}^{x_{k+1}} f(x)\, dx \approx f(x_k)(x_{k+1} - x_k) = f(x_k)\Delta x,$$

and taking the limit as $\Delta x$ tends to zero therefore yields

$$E(X) = \lim_{\Delta x \to 0} \sum_{k=-\infty}^{\infty} x_k f(x_k)\Delta x = \int_{-\infty}^{\infty} x f(x)\, dx. \tag{2.16}$$

**2.5.2 Definition.** If $X : S \to \mathbb{R}$ is a continuous random variable with density $f$, then

$$E(X) := \int_{-\infty}^{\infty} x f(x)\, dx,$$

and the variance of $X$ is

$$\mathrm{Var}(X) := E((X - E(X))^2) = \int_{-\infty}^{\infty} (x - E(X))^2 f(x)\, dx.$$

Since we know that a random variable is discrete whenever its sample space domain is discrete, it follows that a non-discrete random variable must be defined on a sample space that is also non-discrete. Consequently, the present discussion raises the question as to how equation (2.3) can be generalized to the case where $S$ is continuous. To answer this question, we recall that in the calculus sums are routinely turned into integrals as discrete evaluation points $x_k$ give way to evaluation points $x$ that are spread continuously (see, for example, equation (2.16)). And indeed, this the right approach: the sum over the discrete outcomes $s \in S$ in (2.3) is to be replaced by an integral over the continuous set $S$ and the discrete probabilities $P(\{s\})$ assume the form $p(s) \, ds$ with $p$ being a non-negative density function defined on $S$. Hence the continuous-case sample space formula for the mean of $X$ is

$$E(X) = \int_S X(s) p(s) \, ds. \tag{2.17}$$

Moreover, and to clarify the role of $p$, we notice that the probability of any event $E \subset S$ is here understood to be

$$P(E) = \int_E p(s) \, ds, \tag{2.18}$$

and that, in particular,

$$1 = P(S) = \int_S p(s) \, ds.$$

To further elucidate the meaning of these equations, we wish to point out that the variable $s$ is purely generic and that it would, for example, be replaced by a pair of variables, say, $(s, t)$ in the case where $S$ is a region in $\mathbb{R}^2$. So instead of (2.17) we would write

$$E(X) = \int_S X(s, t) p(s, t) \, ds \, dt,$$

and for regions in $\mathbb{R}^3$ the number of variables would naturally be increased to three.

*Remark.* If $s$ is a real number and $X(s) = s$, then the integral in (2.17) is equal to $\int_S s p(s) \, ds$ which is formally very similar to the integral $\int_{-\infty}^{\infty} x f(x) \, dx = \int_{\mathbb{R}} x f(x) \, dx$ in Definition 2.5.2. Consequently, we may rightly wonder whether these two formulas are in fact truly distinct or whether they only appear to be so. As it turns out, the correct answer is, "yes and no," because (2.17) is more general and includes the formula in Definition 2.5.2 as a special case. Most notably, the domain of integration $S$ in (2.17) can assume a wide variety of forms and is not always $\mathbb{R}$ as in the integral in Definition 2.5.2. Furthermore, offering (2.17) as a distinct alternative is helpful insofar as its application in several of the examples below will be very different in its computational character from the application of the formula in Definition 2.5.2 and frequently also quite a bit simpler.

*Remark.* The Theorems 2.3.5, 2.3.7, and 2.3.12, which we previously established for discrete random variables, remain valid as well in the continuous case because all we need to do in adjusting the relevant proofs is to replace sums by integrals. For instance, Theorem 2.3.5a remains valid because

$$E(X + Y) = \int_S (X(s) + Y(s))p(s)\,ds = \int_S X(s)p(s)\,ds + \int_S Y(s)p(s)\,ds$$
$$= E(X) + E(Y).$$

For clarity, we also wish to mention that the integral representations of the terms $E(X^2)$ and $E(e^{tX})$, referred to in Theorems 2.3.7 and 2.3.12, are

$$E(X^2) = \int_{-\infty}^{\infty} x^2 f(x)\,dx = \int_S X(s)^2 p(s)\,ds$$

and

$$E(e^{tX}) = \int_{-\infty}^{\infty} e^{tx} f(x)\,dx = \int_S e^{tX(s)} p(s)\,ds.$$

Furthermore and by extension, the integral version of the general formula (2.4) is

$$E(g(X)) = \int_{-\infty}^{\infty} g(x)f(x)\,dx = \int_S g(X(s))p(s)\,ds.$$

To proceed, we generalize the formula for $f$ in (2.13) by introducing the following definition:

**2.5.3 Definition.** A random variable $X : S \to \mathbb{R}$ is said to be *uniformly distributed* over the interval $[a, b]$ if its density is

$$f(x) = \begin{cases} 1/(b - a) & \text{if } x \in [a, b], \\ 0 & \text{otherwise.} \end{cases}$$

**2.5.4 Example.** If $X$ is uniformly distributed over $[a, b]$, then

$$E(X) = \int_{-\infty}^{\infty} xf(x)\,dx = \int_a^b \frac{x}{b - a}\,dx = \frac{b^2 - a^2}{2(b - a)} = \frac{a + b}{2},$$

and

$$E(X^2) = \int_{-\infty}^{\infty} x^2 f(x)\,dx = \int_a^b \frac{x^2}{b - a}\,dx = \frac{b^3 - a^3}{3(b - a)} = \frac{a^2 + ab + b^2}{3}.$$

Hence

$$\text{Var}(X) = E(X^2) - E(X)^2 = \frac{a^2 + ab + b^2}{3} - \frac{a^2 + 2ab + b^2}{4} = \frac{(a-b)^2}{12}.$$

Note: the fact that $E(X) = (a+b)/2$ makes perfect sense because $(a+b)/2$ is the midpoint of the interval $[a, b]$.

**2.5.5 Example.** Let $X(s) := s^2$ for all $s \in [0, 2]$ and assume that all $s \in [0, 2]$ are equally likely to occur (so $s$ is uniformly distributed over $[0, 2]$). Then $R(X) = [0, 4]$, and for any $x \in [0, 4]$ we have

$$P(X \le x) = P(\{s \in [0, 2] \mid s^2 \le x\}) = P(\{s \in [0, 2] \mid s \le \sqrt{x}\})$$
$$= \frac{\text{length of } [0, \sqrt{x}]}{\text{length of } [0, 2]} = \frac{\sqrt{x}}{2}.$$

Thus

$$F(x) = P(X \le x) = \begin{cases} 0 & \text{if } x < 0, \\ \sqrt{x}/2 & \text{if } 0 \le x \le 4, \\ 1 & \text{if } x > 4, \end{cases}$$

and, by implication,

$$f(x) = F'(x) = \begin{cases} 1/(4\sqrt{x}) & \text{if } 0 \le x \le 4, \\ 0 & \text{otherwise.} \end{cases}$$

Hence

$$E(X) = \int_0^4 \frac{x}{4\sqrt{x}} \, dx = \frac{1}{6}\sqrt{x}^3 \Big|_0^4 = \frac{4}{3}$$

and

$$\text{Var}(X) = E(X^2) - E(X)^2 = \int_0^4 \frac{x^2}{4\sqrt{x}} \, dx - \frac{16}{9} = \frac{32}{10} - \frac{16}{9} = \frac{64}{45}.$$

To confirm these results, we can use integrals over the sample space $S = [0, 2]$ with $p(s) = 1/2$ (because $s$ is uniformly distributed over $[0, 2]$). This yields

$$E(X) = \int_S X(s)p(s) \, ds = \int_0^2 s^2 \frac{ds}{2} = \frac{4}{3}$$

and

$$\text{Var}(X) = \int_0^2 s^4 \frac{ds}{2} - \frac{16}{9} = \frac{32}{10} - \frac{16}{9} = \frac{64}{45},$$

as desired.

**2.5.6 Example.** Let $X(s,t) := st$ for all $(s,t) \in (0,1] \times (0,1] =: S$ and assume that all points $(s,t) \in S$ are equally likely to occur. Then $R(X) = (0,1]$, and for any $x \in (0,1]$ we have

$$P(X \leq x) = P(\{(s,t) \in S \mid st \leq x\}) = P(\{(s,t) \in S \mid t \leq x/s\}).$$

Since the set $F_x := \{(s,t) \in S \mid t \leq x/s\}$ is readily seen to be the shaded region in Figure 2.2 and since the area of this region is equal to the area of a

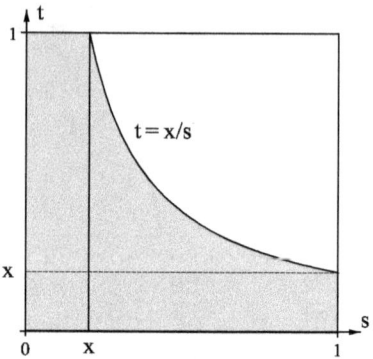

Figure 2.2: the region $F_x$.

rectangle with side lengths 1 and $x$ plus the area under the curve described by the equation $t = x/s$ along the $s$-axis interval $[x,1]$, it follows that

$$P(F_x) = \frac{x + \int_x^1 x/s \, ds}{\text{area of } (0,1] \times (0,1]} = \frac{x - x\ln(x)}{1} = x - x\ln(x).$$

Hence

$$F(x) = P(X \leq x) = \begin{cases} 0 & \text{if } x \leq 0, \\ x - x\ln(x) & \text{if } x \in (0,1], \\ 1 & \text{if } x > 1, \end{cases}$$

and, by implication,

$$f(x) = F'(x) = \begin{cases} -\ln(x) & \text{if } x \in (0,1], \\ 0 & \text{otherwise.} \end{cases}$$

Using this density $f$, we further find that

$$E(X) = \int_{-\infty}^{\infty} x f(x) \, dx = -\int_0^1 x \ln(x) \, dx = -\left(\frac{x^2 \ln(x)}{2} - \frac{x^2}{4}\right)\Big|_0^1 = \frac{1}{4}$$

and

$$E(X^2) = \int_{-\infty}^{\infty} x^2 f(x) \, dx = -\int_0^1 x^2 \ln(x) \, dx = -\left(\frac{x^3 \ln(x)}{3} - \frac{x^3}{9}\right)\Big|_0^1 = \frac{1}{9}.$$

Thus

$$\text{Var}(X) = \frac{1}{9} - \frac{1}{4^2} = \frac{5}{144}.$$

To confirm these results, we use integrals over the sample space with $p(s,t) = 1$ (because all points $(s,t)$ are equally likely to occur and because the total area of the square-shaped sample space $(0,1] \times (0,1]$ is 1). This yields

$$E(X) = \int_S X(s,t) p(s,t) \, ds \, dt = \int_0^1 \int_0^1 st \, ds \, dt = \frac{1}{2} \int_0^1 t \, dt = \frac{1}{4}$$

and

$$E(X^2) = \int_S X(s,t)^2 p(s,t) \, ds \, dt = \int_0^1 \int_0^1 s^2 t^2 \, ds \, dt = \frac{1}{3} \int_0^1 t^2 \, dt = \frac{1}{9},$$

as desired.

**2.5.7 Example.** Let us assume that points $(s,t)$ are chosen with equal probability from within the unit disc $S = \{(s,t) \in \mathbb{R}^2 \mid s^2 + t^2 \leq 1\}$. Then $p(s,t) = 1/\pi$ (because the area of $S$ equals $\pi$), and therefore, using integration in polar coordinates (with $s = r\cos(\theta)$, $t = r\sin(\theta)$, and $ds \, dt = r \, dr \, d\theta$), we find that the expected value of $X(s,t) := s^2 + t^2$ is

$$E(X) = \int_S X(s,t) \, ds \, dt = \int_0^{2\pi} \int_0^1 \frac{r^2}{\pi} r \, dr \, d\theta = 2 \int_0^1 r^3 dr = \frac{1}{2}. \qquad (2.19)$$

Furthermore, since the inequality $s^2 + t^2 \leq x$ describes a disc of radius $\sqrt{x}$ the area of which is $\pi x$ for any $x \geq 0$, the cumulative density of $X$ is

$$F(x) = P(X \leq x) = P(s^2 + t^2 \leq x) = \begin{cases} 0 & \text{if } x < 0, \\ \pi x / \pi = x & \text{if } 0 \leq x \leq 1, \\ 1 & \text{if } x > 1. \end{cases}$$

By implication, the density of $X$ is

$$f(x) = F'(x) = \begin{cases} 1 & \text{if } 0 \leq x \leq 1, \\ 0 & \text{otherwise,} \end{cases}$$

and hence we may confirm (2.19) as follows:

$$E(X) = \int_{-\infty}^{\infty} x f(x) \, dx = \int_0^1 x \, dx = \frac{1}{2}.$$

**2.5.8 Example.** Assume that for a given value $\lambda \in \mathbb{R}$ the density of a random variable $X$ is of the form

$$f(x) = \begin{cases} \lambda e^{-3x} & \text{if } x \geq 0, \\ 0 & \text{if } x < 0. \end{cases}$$

In order to determine $\lambda$, we make use of the fact that the total integral of $f$ must be 1. This yields

$$1 = \int_{-\infty}^{\infty} f(x)\, dx = \int_{0}^{\infty} \lambda e^{-3x}\, dx = -\frac{\lambda e^{-3x}}{3}\Big|_{0}^{\infty} = \frac{\lambda}{3},$$

and, by implication,

$$\lambda = 3.$$

Using this value for $\lambda$, the moment-generating function of $X$ is

$$m_X(t) = E(e^{tX}) = \int_{0}^{\infty} 3e^{tx}e^{-3x}\, dx = -\frac{3e^{-(3-t)x}}{3-t}\Big|_{0}^{\infty} = \frac{3}{3-t},$$

and therefore,

$$E(X) = m_X'(0) = \frac{3}{(3-t)^2}\Big|_{t=0} = \frac{1}{3}.$$

Furthermore, since

$$E(X^2) = m_X''(0) = \frac{6}{(3-t)^3}\Big|_{t=0} = \frac{2}{9},$$

we find that

$$\text{Var}(X) = E(X^2) - E(X)^2 = \frac{2}{9} - \frac{1}{9} = \frac{1}{9}.$$

# Exercises

**2.5.9.** Find the cumulative density and the regular density of the random variable $X(s) := s(4 - s)$ defined on $[0, 4]$ (the density on $[0, 4]$ is assumed to be uniform). Then use your result to find the mean and the variance of $X$.

**2.5.10.** Use integrals over the sample space to verify your results for the mean and the variance in Exercise 2.5.9.

**2.5.11.** What is the probability for the value of the random variable defined in Exercise 2.5.9 to be between 1 and 3?

**2.5.12.** Assume that $X$ is a random variable with density $f(x) = Cx^2$ and range $R(X) = [1, 3]$.

a) Find the value of $C$.

b) Find $F(x)$ for $x \in \mathbb{R}$.

c) Use b) to determine $P(1 \le X \le 2)$.

**2.5.13.** Assume that $X$ is a continuous random variable with range $R(X) = [1, 3]$. What are the values of $F(0)$, $F(1)$, $F(3)$, $F(3.5)$, and $P(2 \le X \le 2)$?

**2.5.14.** Assume that $X$ is a continuous random variable with density

$$f(x) = \begin{cases} 0 & \text{if } x \le 0, \\ 2e^{-\alpha x} & \text{if } x > 0. \end{cases}$$

Find the value of $\alpha$ and determine the cumulative density $F(x)$.

**2.5.15.** Find the cumulative probability density and the regular density of the random variable $X(s, t) := s^2 t$ defined on $[0, 2] \times [0, 2]$ (the density of $(s, t)$ is assumed to be uniform in the sense that all $(s, t)$ are equally likely to occur).

**2.5.16.** Find the mean of the random variable, defined in Exercise 2.5.15, using...

**a)** an integral over the sample space.

**b)** the density function determined in Exercise 2.5.15.

**2.5.17.** What is the probability for the value of the random variable, defined in Exercise 2.5.15, to fall between 4 and 6?

**2.5.18.** Assume that the density of a random variable $X$ is

$$f(x) = \begin{cases} C/x & \text{if } x \in [1, 3], \\ 0 & \text{otherwise.} \end{cases}$$

**a)** Find the value of $C$.

**b)** Determine the probability for $X$ to be greater than 2.

**c)** Find the mean and variance of $X$.

**2.5.19.** Assume that the density of a random variable $X$ is

$$f(x) = \begin{cases} Ce^{-4x} & \text{if } x > 0, \\ 0 & \text{otherwise.} \end{cases}$$

**a)** Find the value of $C$.

**b)** Find the moment-generating function of $X$.

**c)** Find the mean and variance of $X$.

**2.5.20.** Use an integral over the sample space to find the mean and variance of the random variable $X(s) := |s|$ (for $s \in \mathbb{R}$), given that the probability of an event $E \subset \mathbb{R}$ is $P(E) = \int_E f(s) \, ds$ with $f(s) = \lambda e^{-|s|}$. Note: $\lambda$ is a constant that needs to determined in such a way that $P(S) = P(\mathbb{R}) = 1$.

**2.5.21.** Find the cumulative density of $X$ in Exercise 2.5.20, and sketch the graph of this cumulative density.

**2.5.22.** Let $X(s,t) := \sqrt{s^2 + t^2}$ for $(s,t) \in S := \mathbb{R}^2$, and assume that the probability of an event $E \subset \mathbb{R}^2$ is $P(E) = \int_E f(s,t) \, dA$ with $f(s,t) = \lambda e^{-\sqrt{s^2+t^2}}$ (here again $\lambda$ needs to be determined from the assumption that $P(S) = 1$). Use an integral over the sample space to find the mean and variance of $X$.

**2.5.23.** Find the cumulative density of $X$ in Exercise 2.5.22.

## 2.6 The Normal Distribution

Suppose that you are throwing darts at an infinite horizontal $st$-plane with a central target at the origin. Each dart hits the plane at some point $(s,t)$, and the sample space representing these hits is therefore $\mathbb{R}^2 = \{(s,t) \mid s,t \in \mathbb{R}\}$. Furthermore, in denoting by $P(E)$ the probability for a dart to hit a region $E \subset \mathbb{R}^2$, we may appeal to equation (2.18) to conjecture that there ought to be a function $p$ defined on $S = \mathbb{R}^2$ such that

$$P(E) = \int_E p(s,t) \, ds \, dt. \tag{2.20}$$

In order to find a formula for $p$, we need to think carefully about the properties that we can expect $p$ to satisfy. For instance, it certainly is natural to assume that the density of hits in the plane depends only on the distance from the central target at the origin and not on the direction relative to this target. In other words, we may expect that missing the target by a certain margin in one direction is just as likely as missing the target by the same margin in any other direction. Consequently, instead of an arbitrary dependence on $s$ and $t$, we may expect $p$ to depend only on the distance $\sqrt{s^2 + t^2}$ or, equivalently, on $s^2 + t^2$. So there ought to be a function $h : \mathbb{R} \to \mathbb{R}$ such that

$$p(s,t) = h(s^2 + t^2). \tag{2.21}$$

To further analyze the properties of $p$, it is helpful to set $S := \mathbb{R}^2$ and to define a random variable $X : S \to \mathbb{R}$ via the equation $X(s,t) := s$. Given this definition, the cumulative density of $X$ is

$$F(x) = P(X \le x) = \int_{-\infty}^{x} \int_{-\infty}^{\infty} p(s,t) \, dt \, ds,$$

and the corresponding regular density is

$$f(x) = F'(x) = \int_{-\infty}^{\infty} p(x,t) \, dt.$$

Since (2.21) evidently implies that $p(s,t) = p(t,s)$ for all $(s,t) \in \mathbb{R}^2$, it follows that the density of $Y(s,t) := t$ is equal to $f$ as well:

$$\int_{-\infty}^{\infty} p(s,y) \, ds = \int_{-\infty}^{\infty} p(y,s) \, ds = f(y).$$

Furthermore, since there is no reason to assume that the $s$ and $t$-coordinates are in any way dependent on each other in the sense that knowledge of one coordinate implies knowledge of the other, we may suppose that the events

$$E_x := \{X \leq x\} = \{(s,t) \in \mathbb{R}^2 \mid s \leq x\}$$

and

$$F_y := \{Y \leq y\} = \{(s,t) \in \mathbb{R}^2 \mid t \leq y\}$$

are independent. That is to say, it ought to be the case that

$$P(E_x \cap F_y) = P(E_x)P(F_y). \tag{2.22}$$

Since $P(E_x)$ is the probability for $X$ to be less than or equal to $x$, it follows that

$$P(E_x) = \int_{-\infty}^{x} f(u)\,du,$$

and similarly,

$$P(F_y) = \int_{-\infty}^{y} f(v)\,dv.$$

Moreover, since $E_x \cap E_y = \{(s,t) \in \mathbb{R}^2 \mid s \leq x \wedge t \leq y\}$, we have

$$P(E_x \cap E_y) = \int_{-\infty}^{x} \int_{-\infty}^{y} p(s,t)\,dt\,ds,$$

and therefore, (2.22) implies that

$$\int_{-\infty}^{x} \int_{-\infty}^{y} p(s,t)\,dt\,ds = \int_{-\infty}^{x} f(u)\,du \int_{-\infty}^{y} f(v)\,dv.$$

Taking the partial derivative with respect to $x$ yields

$$\int_{-\infty}^{y} p(x,t)\,dt = f(x) \int_{-\infty}^{y} f(v)\,dv,$$

and taking the partial derivative with respect to $y$ in this resulting equation yields

$$p(x,y) = f(x)f(y). \tag{2.23}$$

To proceed, we combine (2.21) with (2.23) to conclude that

$$h(x^2 + y^2) = f(x)f(y),$$

and taking the partial derivatives with respect to $x$ and $y$, we find that

$$2xh'(x^2 + y^2) = \frac{\partial}{\partial x}h(x^2 + y^2) = \frac{\partial}{\partial x}(f(x)f(y)) = f'(x)f(y),$$

$$2yh'(x^2 + y^2) = \frac{\partial}{\partial y}h(x^2 + y^2) = \frac{\partial}{\partial y}(f(x)f(y)) = f(x)f'(y).$$

Solving both of these equations for $h'(x^2 + y^2)$ yields

$$\frac{f'(x)f(y)}{2x} = h'(x^2 + y^2) = \frac{f(x)f'(y)}{2y},$$

and, by implication,

$$\frac{f'(x)}{2xf(x)} = \frac{f'(y)}{2yf(y)}. \tag{2.24}$$

Since the left-hand side of this equation depends only on $x$ and the right-hand side only on $y$, it follows that both sides are actually constant. So there is a $C \in \mathbb{R}$ such that

$$\frac{f'(x)}{f(x)} = 2Cx.$$

Integrating both sides of this equation, it is easy to see that there is a constant $K \in \mathbb{R}$ such that

$$\ln |f(x)| = Cx^2 + K,$$

and therefore, with $D := \pm e^K$, we find that

$$f(x) = De^{Cx^2}. \tag{2.25}$$

Consequently, (2.23) implies that

$$p(s, t) = D^2 e^{C(s^2 + t^2)}. \tag{2.26}$$

Since the probability for the occurrence of dart hits in $\mathbb{R}^2$ may naturally be assumed to be decreasing away from the origin, the constant $C$ must be less than zero, and in setting $\sigma := \sqrt{-1/(2C)}$, we may therefore rewrite (2.26) as follows:

$$p(s, t) = D^2 e^{-(s^2 + t^2)/2\sigma^2}. \tag{2.27}$$

Using equation (2.20) with $S = \mathbb{R}^2$ in place of $E$, we obtain

$$1 = P(S) = \int_S p(s, t) \, ds \, dt = \int_{-\infty}^{\infty} \int_{-\infty}^{\infty} D^2 e^{-(s^2 + t^2)/2\sigma^2} \, ds \, dt,$$

and in transforming the integral on the right into polar coordinates, we find that

$$1 = \int_0^{2\pi} \int_0^{\infty} D^2 e^{-r^2/2\sigma^2} r \, dr \, d\theta = \int_0^{2\pi} -\sigma^2 D^2 e^{-r^2/2\sigma^2} \Big|_0^{\infty} d\theta$$

$$= \int_0^{2\pi} \sigma^2 D^2 \, d\theta = 2\pi D^2 \sigma^2.$$

Hence

$$D = \frac{1}{\sqrt{2\pi}\sigma},$$

and substituting this value for $D$ in (2.25) and (2.27) finally yields

$$p(s,t) = \frac{e^{-(s^2+t^2)/2\sigma^2}}{2\pi\sigma^2} \quad \text{and} \quad f(x) = \frac{e^{-x^2/2\sigma^2}}{\sqrt{2\pi}\sigma}. \tag{2.28}$$

**2.6.1 Example.** In Figure 2.3 we are shown two graphs of $p$ corresponding to the values $\sigma = 1$ and $\sigma = 2$. These graphs illustrate that for smaller values

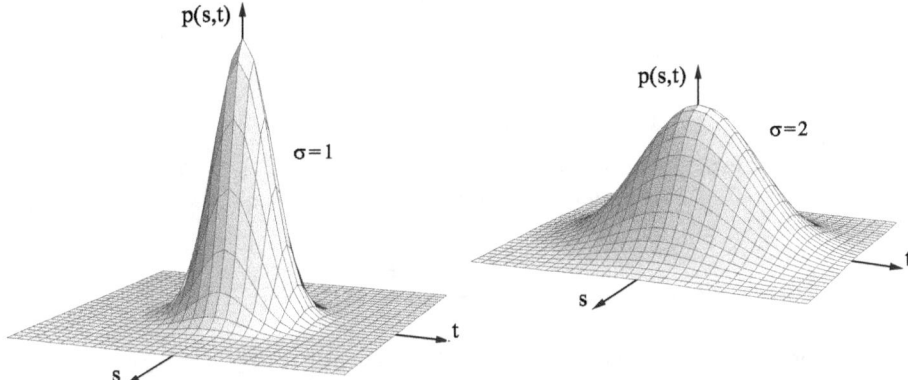

Figure 2.3: graphs of $p$ for $\sigma = 1$ and $\sigma = 2$.

of $\sigma$ the probability density is less spread out in the plane than for larger ones. Intuitively speaking, this difference in shape bespeaks different levels of skill in throwing darts, because the smaller spread on the left indicates a higher concentration of dart hits close to the central target at $(0,0)$. A similar dependence of the density spread on $\sigma$ we also observe in the distribution of the $s$- or $x$-coordinates of hits, as described by the probability density $f$ in (2.28). In Figure 2.4 we see that for smaller values of $\sigma$ the probability density $f$ is more narrowly concentrated at the origin than for larger ones.

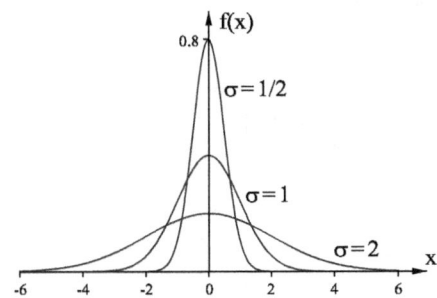

Figure 2.4: the probability density $f$ for $\sigma = 1/2$, 1, and 2.

**2.6.2 Example.** Let us determine the probability that a dart hits the plane at a distance less than or equal to 2 and greater than or equal to 1 under the assumption that $\sigma = 1$. In other words, we wish to determine the probability $P(A)$ for the annulus $A$ with inner radius 1 and outer radius 2 that is shown in Figure 2.5. Using polar coordinates to evaluate the integral in (2.20) with $A$ in place of $E$, we obtain

$$P(A) = \int_A p(s,t)\,ds\,dt = \int_0^{2\pi}\int_1^2 \frac{e^{-r^2/2}}{2\pi}\,r\,dr\,d\theta = \int_0^{2\pi} \left. -\frac{e^{-r^2/2}}{2\pi}\right|_1^2 d\theta$$

$$= \int_0^{2\pi} \frac{e^{-1/2} - e^{-2}}{2\pi}\,d\theta = e^{-1/2} - e^{-2} \approx 0.471.$$

Motivated by the results in (2.28), we introduce the following definition:

**2.6.3 Definition.** A multivariable function $f : \mathbb{R}^n \to \mathbb{R}$ is said to be a *normal (or Gaussian) probability density* if there are constants $\sigma > 0$ and $\mu_1, \ldots, \mu_n \in \mathbb{R}$ such that

$$f(x_1, \ldots, x_n) = \frac{e^{-((x_1-\mu_1)^2+\cdots+(x_n-\mu_n)^2)/2\sigma^2}}{\sqrt{2\pi}^n \sigma^n}. \tag{2.29}$$

Furthermore, a random variable $X : S \to \mathbb{R}$ is said to be *normal* or *normally distributed* if there are constants $\sigma > 0$ and $\mu \in \mathbb{R}$ such that the density of $X$ is the *normal density*

$$f(x) = \frac{e^{-(x-\mu)^2/2\sigma^2}}{\sqrt{2\pi}\sigma}. \tag{2.30}$$

In particular, if $\mu = 0$ and $\sigma = 1$, this density is said to be *standard normal* and will be denoted by $z(x)$, that is,

$$z(x) = \frac{e^{-x^2/2}}{\sqrt{2\pi}}.$$

Note: for the normal probability densities given in (2.28) all the $\mu$-values in (2.29) and (2.30) are equal to zero.

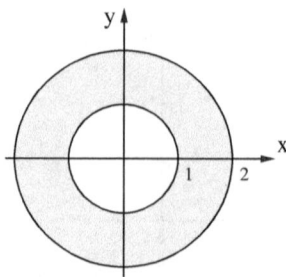

Figure 2.5: the annulus $A$.

In order to verify that the total probabilities described by the density functions in Definition 2.6.3 are equal to one, we argue as follows: from the preceding discussion we know that the integral of $p(s, t)$ over $\mathbb{R}^2$ is equal to one. Hence

$$1 = \int_{\mathbb{R}^2} p(s, t)\, ds\, dt = \int_{-\infty}^{\infty} \int_{-\infty}^{\infty} \frac{e^{-(s^2+t^2)/2\sigma^2}}{2\pi\sigma^2}\, ds\, dt$$

$$= \int_{-\infty}^{\infty} \int_{-\infty}^{\infty} \frac{e^{-s^2/2\sigma^2}}{\sqrt{2\pi}\sigma} \cdot \frac{e^{-t^2/2\sigma^2}}{\sqrt{2\pi}\sigma}\, ds\, dt$$

$$= \int_{-\infty}^{\infty} \frac{e^{-s^2/2\sigma^2}}{\sqrt{2\pi}\sigma}\, ds \int_{-\infty}^{\infty} \frac{e^{-t^2/2\sigma^2}}{\sqrt{2\pi}\sigma}\, dt = \left( \int_{-\infty}^{\infty} \frac{e^{-u^2/2\sigma^2}}{\sqrt{2\pi}\sigma}\, du \right)^2,$$

and therefore, the substitution $u := x - \mu$ yields

$$\int_{-\infty}^{\infty} \frac{e^{-(x-\mu)^2/2\sigma^2}}{\sqrt{2\pi}\sigma}\, dx = \int_{-\infty}^{\infty} \frac{e^{-u^2/2\sigma^2}}{\sqrt{2\pi}\sigma}\, du = 1,$$

and this result in turn allows us to infer that

$$\int_{\mathbb{R}^n} \frac{e^{-((x_1-\mu_1)^2+\cdots+(x_n-\mu_n)^2)/2\sigma^2}}{\sqrt{2\pi}^n \sigma^n}\, dx_1 \ldots dx_n$$

$$= \int_{-\infty}^{\infty} \cdots \int_{-\infty}^{\infty} \prod_{k=1}^{n} \frac{e^{-(x_k-\mu_k)^2/2\sigma^2}}{\sqrt{2\pi}\sigma}\, dx_1 \ldots dx_n$$

$$= \prod_{k=1}^{n} \int_{-\infty}^{\infty} \frac{e^{-(x_k-\mu_k)^2/2\sigma^2}}{\sqrt{2\pi}\sigma}\, dx_k = 1,$$

as desired.

**2.6.4 Theorem.** *If $f$ is a normal density on $\mathbb{R}$, as given in (2.30), then $f$ has a global maximum at $\mu$, and $f$ has points of inflection at $\mu \pm \sigma$ (see Figure 2.6 for a visual illustration and see Exercise 2.6.13 for the proof).*

**2.6.5 Theorem.** *If $X$ is a normal random variable (with parameters $\mu$ and $\sigma$), then*

   a) $m_X(t) = e^{\mu t + \sigma^2 t^2/2}$,

   b) $E(X) = \mu$, and

   c) $\mathrm{Var}(X) = \sigma^2$.

*Proof.* Since $m_X(t) = E(e^{tX})$, it follows that

$$m_X(t) = \int_{-\infty}^{\infty} \frac{e^{tx} e^{-(x-\mu)^2/2\sigma^2}}{\sqrt{2\pi}\sigma}\, dx = \int_{-\infty}^{\infty} \frac{e^{-(x^2-2(\mu+\sigma^2 t)x+\mu^2)/2\sigma^2}}{\sqrt{2\pi}\sigma}\, dx$$

$$= \int_{-\infty}^{\infty} \frac{e^{-((x-(\mu+\sigma^2 t))^2-(\mu+\sigma^2 t)^2+\mu^2)/2\sigma^2}}{\sqrt{2\pi}\sigma}\, dx$$

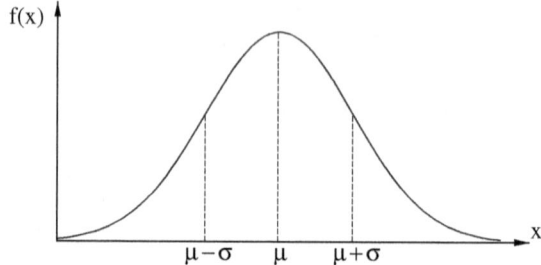

Figure 2.6: a normal probability density on $\mathbb{R}$.

$$= e^{\mu t + \sigma^2 t^2/2} \int_{-\infty}^{\infty} \frac{e^{-(x-(\mu+\sigma^2 t))^2/2\sigma^2}}{\sqrt{2\pi}\sigma}\, dx = e^{\mu t + \sigma^2 t^2/2}.$$

Hence

$$E(X) = m_X'(0) = (\mu + \sigma^2 t)e^{\mu t + \sigma^2 t^2/2}\Big|_{t=0} = \mu$$

and

$$\mathrm{Var}(X) = E(X^2) - E(X)^2 = m_X''(0) - E(X)^2$$
$$= (\sigma^2 + (\mu + \sigma^2 t)^2)e^{\mu t + \sigma^2 t^2/2}\Big|_{t=0} - \mu^2 = \sigma^2 + \mu^2 - \mu^2 = \sigma^2,$$

as desired.                                                                  □

The purpose of the next theorem is to establish the convenient and useful fact that every normal integral can be turned into a standard normal integral and vice versa.

**2.6.6 Theorem.** *If $X$ is a normal random variable with mean $\mu$ and standard deviation $\sigma$ (see Theorem 2.6.5), then*

$$P(\mu + a\sigma \le X \le \mu + b\sigma) = \int_a^b \frac{e^{-x^2/2}}{\sqrt{2\pi}}\, dx = \int_a^b z(x)\, dx$$

*and*

$$P(a \le X \le b) = \int_{(a-\mu)/\sigma}^{(b-\mu)/\sigma} \frac{e^{-x^2/2}}{\sqrt{2\pi}}\, dx = \int_{(a-\mu)/\sigma}^{(b-\mu)/\sigma} z(x)\, dx$$

*for all $a, b \in \mathbb{R}$ with $a < b$.*

*Proof.* Introducing the substitution $u = (x - \mu)/\sigma$, we find that $du = dx/\sigma$, and therefore,

$$P(\mu + a\sigma \le X \le \mu + b\sigma) = \int_{\mu+a\sigma}^{\mu+b\sigma} \frac{e^{-(x-\mu)^2/2\sigma^2}}{\sqrt{2\pi}\sigma}\, dx$$

$$= \int_a^b \frac{e^{-u^2/2}}{\sqrt{2\pi}}\, du = \int_a^b \frac{e^{-x^2/2}}{\sqrt{2\pi}}\, dx.$$

Replacing in this equation $a$ by $(a - \mu)/\sigma$ and $b$ by $(b - \mu)/\sigma$, it follows that

$$P(a \leq X \leq b) = \int_{(a-\mu)/\sigma}^{(b-\mu)/\sigma} \frac{e^{-x^2/2}}{\sqrt{2\pi}} \, dx,$$

as desired. $\qquad\square$

**2.6.7 Corollary.** *If $X$ is a normal random variable with mean $\mu$ and standard deviation $\sigma$, then*

$$\frac{X - \mu}{\sigma}$$

*is standard normal.*

*Proof.* We need to show that

$$P\left(a \leq \frac{X - \mu}{\sigma} \leq b\right) = \int_a^b z(x) \, dx$$

for all $a, b \in \mathbb{R}$. Using elementary algebra in conjunction with Theorem 2.6.6, we find that

$$P\left(a \leq \frac{X - \mu}{\sigma} \leq b\right) = P\left(\mu + a\sigma \leq X \leq \mu + b\sigma\right) = \int_a^b z(x) \, dx,$$

as desired. $\qquad\square$

**2.6.8 Corollary.** *(The 68-95-99.7 Rule) If $X$ is a normal random variable with mean $\mu$ and standard deviation $\sigma$, then*

**a)** $P(|X - \mu| \leq \sigma) \approx 0.6827 \approx 0.68$,

**b)** $P(|X - \mu| \leq 2\sigma) \approx 0.9545 \approx 0.95$,

**c)** $P(|X - \mu| \leq 3\sigma) \approx 0.9973 \approx 0.997$.

*Proof.* We will only prove b) as the proofs of a) and c) are completely analogous. Using a computer or calculator along with the first equation in Theorem 2.6.6 yields

$$P(|X - \mu| \leq 2\sigma) = P(\mu - 2\sigma \leq X \leq \mu + 2\sigma) = \int_{-2}^2 \frac{e^{-x^2/2}}{\sqrt{2\pi}} \, dx \approx 0.9545,$$

as desired. $\qquad\square$

The 68-95-99.7 rule is useful primarily because the three values to which it refers—68%, 95%, and 99.7%—are universally applicable to all normal distributions no matter what the values of $\sigma > 0$ and $\mu \in \mathbb{R}$ happen to be. A visual illustration of this rule is provided in Figure 2.7 (where 'A' stands for 'area').

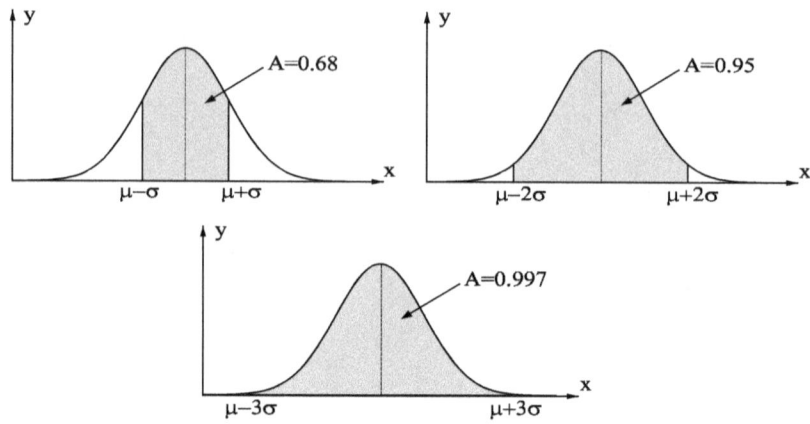

Figure 2.7: the 68-95-99.7 rule.

**2.6.9 Example.** Suppose that the distribution of the bodily height of American males above the age of 18 is approximately normal with mean $\mu = 70$ inches and standard deviation $\sigma = 2.5$ inches. Given this information, we wish to answer the following question: what approximate percentage of American males above the age of 18 have a height greater than 67.5 inches and less than 72.5 inches? Since $67.5 = 70 - 2.5 = \mu - \sigma$ and $72.5 = \mu + \sigma$, the 68-95-99.7 rule implies that the percentage in question is approximately 68. Alternatively, we can also calculate the so-called $z$-score

$$\boxed{\frac{x - \mu}{\sigma}}$$

which tells us how many standard deviations a given value $x$ is away from the mean $\mu$ in either the positive or negative direction. (Note: the name '$z$-score' derives from the fact that a standard normal distribution is commonly also referred to as a $z$-distribution and that we use the substitution $u = (x - \mu)/\sigma$ (see the proof of Theorem 2.6.6) in transforming a normal integral into a standard normal integral.) Using the values 67.5 and 72.5 for $x$, we find that

$$\frac{67.5 - 70}{2.5} = -1$$

and

$$\frac{72.5 - 70}{2.5} = 1,$$

and a distance from the mean of one standard deviation in the positive and negative direction corresponds to the 68% case.

Using the concept of the $z$-score, as introduced in the preceding example, we may restate the 68-95-99.7 rule as follows: if a random variable $X$ (describing, for instance, a person's bodily height) is normally distributed with mean $\mu$ and standard deviation $\sigma$, then the probability for a randomly generated output value $x$ of $X$ to fall between the values $a$ and $b$ is approximately equal to...

| |
|---|
| ... 0.68 if the $z$-score of $a$ is $-1$ and the $z$-score of $b$ is 1. |
| ... 0.95 if the $z$-score of $a$ is $-2$ and the $z$-score of $b$ is 2. |
| ... 0.997 if the $z$-score of $a$ is $-3$ and the $z$-score of $b$ is 3. |

$$(2.31)$$

**2.6.10 Example.** We wish to determine the probability for a randomly selected American male above the age of 18 to have a height between $a = 72.5$ inches and $b = 75$ inches. Since the $z$-scores of $a$ and $b$ are

$$\frac{a - \mu}{\sigma} = \frac{72.5 - 70}{2.5} = 1$$

and

$$\frac{b - \mu}{\sigma} = \frac{75 - 70}{2.5} = 2,$$

respectively, the symmetry of the normal density curve in conjunction with the 68-95-99.7 rule, as stated in (2.31), implies that the probability in question is approximately equal to

$$\frac{0.95 - 0.68}{2} = 0.135.$$

In other words, approximately 13.5% of American males above the age of 18 have a height between 72.5 and 75 inches.

In generalizing the result of Example 2.6.10, we readily find the following dependencies between the probability that a normal random variable $X$ is greater than $a$ and less than $b$ on the one hand and the $z$-scores of $a$ and $b$ on the other:

| $z$-score of $a$ | $z$-score of $b$ | $P(a \leq X \leq b) \approx$ |
|---|---|---|
| $-3$ | $-2$ | $(0.997 - 0.95)/2 = 0.0235$ |
| $-2$ | $-1$ | $(0.95 - 0.68)/2 = 0.135$ |
| $-1$ | $0$ | $0.68/2 = 0.34$ |
| $0$ | $1$ | $0.68/2 = 0.34$ |
| $1$ | $2$ | $(0.95 - 0.68)/2 = 0.135$ |
| $2$ | $3$ | $(0.997 - 0.95)/2 = 0.0235$ |

Furthermore, if the $z$-scores of $a$ and $b$ are $-3$ and 3, respectively, then

$$P(X \leq a) = P(X \geq b) \approx \frac{1 - 0.997}{2} = 0.0015. \qquad (2.32)$$

Visually, the content of the table above is displayed in Figure 2.8.

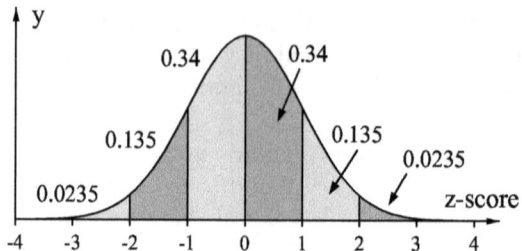

Figure 2.8: approximate areas over successive $z$-score intervals.

*Remark.* The probabilities displayed in the table above and in (2.32) can be made more precise by using the more accurate values in Corollary 2.6.8. For instance, if the $z$-scores of $a$ and $b$ are 1 and 2, respectively, then

$$P(a \leq X \leq b) \approx \frac{0.9545 - 0.6827}{2} = \frac{0.2718}{2} = 0.1359.$$

**2.6.11 Example.** Assume that a random variable $X$ is normally distributed with mean $\mu = 30$ and standard deviation $\sigma = 4$. Then the $z$-score of $a = 38$ is

$$\frac{a - \mu}{\sigma} = \frac{38 - 30}{4} = 2,$$

and the $z$-score of $b = 42$ is

$$\frac{b - \mu}{\sigma} = \frac{42 - 30}{4} = 3.$$

Consequently, according to the table above, the probability for $X$ to lie between 38 and 42 is approximately equal to 0.0235.

## Exercises

**2.6.12.** Find the probability that a dart will hit the plane at a point $(s, t)$ with $s, t \geq 0$ and $s^2 + t^2 \leq 2$ under the assumption that the 'skill parameter' is $\sigma = 2/3$.

**2.6.13.** Use elementary calculus to prove Theorem 2.6.4.

**2.6.14.** Assume that the bodily heights of American males above the age of 18 are distributed approximately normally with mean $\mu = 70$ inches and standard deviation $\sigma = 2.5$ inches. Use the 68-95-99.7 rule to find the approximate percentage of American males above the age of 18 that are taller than 67.5 inches and shorter than 75 inches.

**2.6.15.** A probability exam is taken by a large number of students at a certain college. If the exam scores are distributed approximately normally with overall mean 68% and standard deviation 8.5%, then what percentage of all the students who took the exam had scores greater than or equal to 74% and what percentage had scores less than 85%? Use standard normal integrals to answer these questions.

**2.6.16.** Assume that scores on an intelligence test for the 25 to 35 age group are distributed approximately normally with mean 105 and standard deviation 15 and that scores on the same test for the 60 to 70 age group are distributed normally as well with $\mu = 95$ and $\sigma = 10$. A certain lady, who is 31 years old, scores 120 whereas her father, who is 62 years old, scores 115. Use the 68-95-99.7 rule to determine the percentiles of these scores within the two respective age groups.

**2.6.17.** Assume that $X$ is a normal random variable with parameters $\mu = 60$ and $\sigma = 4$. Use the 68-95-99.7 rule to find approximate values for the following probabilities:

   **a)** $P(|X - 60| \leq 8)$,

   **b)** $P(|X - 60| \geq 12)$,

   **c)** $P(56 \leq X \leq 68)$,

   **d)** $P(64 \leq X \leq 68)$,

   **e)** $P(X \leq 64)$,

   **f)** $P(48 \leq X)$.

**2.6.18.** Express each of the probabilities listed in Exercise 2.6.17 in the form of a standard normal integral.

**2.6.19.** Assume that $X$ is a normal random variable with parameters $\mu = 50$ and $\sigma = 10$. Use standard normal integrals to compute the following probabilities:

   **a)** $P(|X - 50| \leq 5)$,

   **b)** $P(|X - 50| \geq 8)$,

   **c)** $P(56 \leq X \leq 65)$,

   **d)** $P(X \leq 37)$.

## 2.7 The Normal Approximation to the Binomial Distribution

The normal distribution, without any doubt, is far and away the most important distribution in all of probability. And here is why: whenever we produce random values independently in a trial-by-trial identical physical or algorithmic fashion, the sums or averages of these values will be distributed, not normally, but *approximately normally*. Whether we toss a coin or roll a die or count microorganisms in samples of lake water—it doesn't matter—the probabilities for the pertinent sums and averages to lie within a given $z$-score interval can unfailingly be approximated by standard normal integrals. This is truly astonishing and truly of utmost significance insofar as the theory of probability is concerned.

The mathematical theorem that underlies and makes precise this broad applicability of the normal distribution—the so-called *Central Limit Theorem* (Section 3.4)—is a deep result that we don't have the means to rigorously establish in this introductory text. But what we can do and will do in this present section is to discuss the special case of the normal approximation to the binomial distribution and to provide a semi-rigorous outline of a proof that is based on well known calculus techniques and estimates.

To begin with, let us recall that the probability for receiving $k$ heads after $n$ tosses of an unbiased coin is

$$P_n(k) = \frac{1}{2^n} \binom{n}{k} = \frac{1}{2^n} \cdot \frac{n!}{k!(n-k)!}.$$

Considering as an example the case where $n$ equals 40 and plotting the value of $P_{40}(k)$ for $k \in \{0, \ldots, 40\}$ (Figure 2.9), we notice that indeed the resulting graph looks suspiciously similar to the bell-shaped curves in Figures 2.4 and 2.6. Consequently, it appears plausible to conjecture that in general a binomial

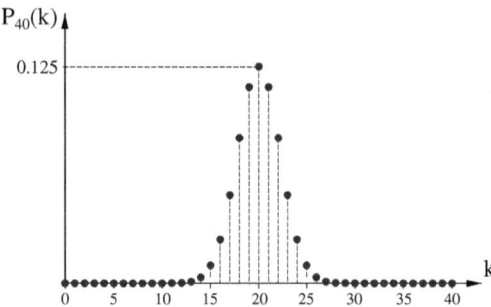

Figure 2.9: plotting $P_{40}(k)$ for $0 \le k \le 40$.

probability density with parameters $n$ and $p$ can be modeled approximately by a normal probability density

$$f_{n,p}(x) := \frac{e^{-(x-\mu_{n,p})^2/2\sigma_{n,p}^2}}{\sqrt{2\pi}\sigma_{n,p}}$$

with

$$\mu_{n,p} := np \quad \text{and} \quad \sigma_{n,p} := \sqrt{np(1-p)}.$$

**2.7.1 Theorem.** *Let $n$ be a positive integer and let $a$ and $b$ be real numbers such that $a < b$. If we set $a_{n,p} := \sigma_{n,p}a + \mu_{n,p}$ and $b_{n,p} := \sigma_{n,p}b + \mu_{n,p}$, then*

$$\int_{a_n}^{b_n} f_{n,p}(x)\, dx = \int_a^b \frac{e^{-x^2/2}}{\sqrt{2\pi}}\, dx \tag{2.33}$$

*and*

$$\lim_{n \to \infty} \sum_{\substack{k=0 \\ a_{n,p} \le k \le b_{n,p}}}^{n} \binom{n}{k} p^k (1-p)^{n-k} = \int_a^b \frac{e^{-x^2/2}}{\sqrt{2\pi}}\, dx. \tag{2.34}$$

In oder to better understand the meaning of this theorem, it is helpful to observe that the $z$-score of $a_{n,p}$ is

$$\frac{a_{n,p} - \mu_{n,p}}{\sigma_{n,p}} = a$$

and that, similarly, the $z$-score of $b_{n,p}$ is $b$. So whenever the $z$-scores of the summation bounds in the binomial distribution are kept constant as $n$ increases to infinity, the corresponding binomial probabilities converge to the standard normal integral the bounds of which are equal to precisely these $z$-scores.

**2.7.2 Example.** Let us use Theorem 2.7.1 to obtain an estimate for the probability that 400 tosses of an unbiased coin produce at least 190, but no more than 205 heads. Since in this case $n = 400$ and $p = 0.5$, it follows that

$$\mu_{400,0.5} = 400 \cdot 0.5 = 200$$

and

$$\sigma_{400,0.5} = \sqrt{400 \cdot 0.5 \cdot 0.5} = 10.$$

Furthermore, since $190 = a_{400,0.5} = 10a + 200$ and $205 = b_{400,0.5} = 10b + 200$, we may infer that $a = -1$ and $b = 1/2$. Using a scientific calculator, we therefore find the estimate

$$\sum_{k=190}^{205} P_{400}(k) \approx \int_{190}^{205} \frac{e^{-(x-200)^2/200}}{10\sqrt{2\pi}}\, dx = \int_{-1}^{1/2} \frac{e^{-x^2/2}}{\sqrt{2\pi}}\, dx \approx 0.53.$$

The error in this estimate is about 6% because a direct evaluation of the sum on the left yields the approximate value 0.561959. As it turns out, this error can be reduced if we interpret the sum above as a midpoint approximation for the area under the graph of $f_{400,0.5}$ over the interval from 189.5 to 205.5. To see why this is so, we need to recall that in the calculus the general form of a midpoint sum over an interval $[a, b]$ is

$$\sum_{k=1}^{n} f\left(\frac{x_{k-1} + x_k}{2}\right) \Delta x,$$

with $\Delta x = (b-a)/n$ and $x_k = a + k\Delta x$ for all $k \in \{0, \ldots, n\}$. Consequently, if this sum is to be equal to the sum of $P_{400}(k)$ from $k = 190$ to 205, then the evaluation points $(x_{k-1} + x_k)/2$ must range through the 16 integers from 190 to 205 as $k$ ranges from 1 to 16. And in order to satisfy this latter requirement, we need to set $\Delta x := (205.5 - 189.5)/16 = 1$ and, correspondingly,

$$x_0 := 189.5, \ x_1 := 190.5, \ x_2 := 191.5, \ldots, \ x_k := 189.5 + k, \ldots, \ x_{16} = 205.5.$$

For given these values $x_k$, the midpoint-sum evaluation points are

$$\frac{x_0 + x_1}{2} = 190, \ \frac{x_1 + x_2}{2} = 191, \ldots, \ \frac{x_{15} + x_{16}}{2} = 205,$$

as desired. As we now re-compute $a$ and $b$ from the adjusted equations $189.5 = 10a + 200$ and $205.5 = 10b + 200$, we find that $a = -1.05$ and $b = 0.55$, and that, by implication,

$$\sum_{k=190}^{205} P_{400}(k) \approx \int_{189.5}^{205.5} \frac{e^{-(x-200)^2/200}}{10\sqrt{2\pi}} \, dx = \int_{-1.05}^{0.55} \frac{e^{-x^2/2}}{\sqrt{2\pi}} \, dx \approx 0.561981.$$

So this time the error is only about 0.004%.

To outline a proof of Theorem 2.7.1, we begin by observing that equation (2.33) is a direct consequence of the first equation in Theorem 2.6.6. Thus we only need to establish (semi-rigorously) the validity of equation (2.34). To do so we recall to begin with the following limit formula which is known as Stirling's formula and which will be established later in this text in the remark on p.205:

$$\lim_{n \to \infty} \frac{n! e^n}{n^n \sqrt{n}} = \sqrt{2\pi}.$$

Using this formula, we find that for large values of $n$, $k$, and $n - k$ it is the case that

$$n! \approx \frac{\sqrt{2\pi} \, n^n \sqrt{n}}{e^n},$$

$$k! \approx \frac{\sqrt{2\pi} \, k^k \sqrt{k}}{e^k}$$

and

$$(n - k)! \approx \frac{\sqrt{2\pi}(n-k)^{n-k}\sqrt{n-k}}{e^{n-k}}.$$

Consequently, in setting

$$f(x) := \frac{p^x (1-p)^{1-x}}{x^x (1-x)^{1-x}} \quad \text{and} \quad g(x) := \frac{1}{\sqrt{x}\sqrt{1-x}},$$

we find that

$$\binom{n}{k} p^k (1-p)^{n-k} \approx \frac{\sqrt{2\pi}\, n^n \sqrt{n}\, e^k e^{n-k} p^k (1-p)^{n-k}}{\sqrt{2\pi}\, k^k \sqrt{k} \sqrt{2\pi}(n-k)^{n-k} \sqrt{n-k}\, e^n}$$

$$= \frac{1}{\sqrt{2\pi}\sqrt{n}} \left( \frac{p^{k/n}(1-p)^{1-k/n}}{(k/n)^{k/n}(1-(k/n))^{1-k/n}} \right)^n \frac{1}{\sqrt{k/n}\sqrt{1-k/n}}$$

$$= \frac{f(k/n)^n g(k/n)}{\sqrt{2\pi}\sqrt{n}}.$$

Hence

$$\sum_{\substack{k=0 \\ a_{n,p} \le k \le b_{n,p}}}^{n} \binom{n}{k} p^k (1-p)^{n-k} \approx \frac{\sqrt{n}}{\sqrt{2\pi}} \sum_{\substack{k=0 \\ a_{n,p} \le k \le b_{n,p}}}^{n} \frac{f(k/n)^n g(k/n)}{n}$$

$$\approx \frac{\sqrt{n}}{\sqrt{2\pi}} \int_{a_{n,p}/n}^{b_{n,p}/n} f(x)^n g(x)\, dx,$$

and therefore, the substitution $x = p + \sqrt{p(1-p)}u/\sqrt{n}$ yields

$$\sum_{\substack{k=0 \\ a_{n,p} \le k \le b_{n,p}}}^{n} \binom{n}{k} p^k (1-p)^{n-k} \approx$$

$$\approx \frac{\sqrt{p(1-p)}}{\sqrt{2\pi}} \int_a^b f\left(p + \frac{\sqrt{p(1-p)}u}{\sqrt{n}}\right)^n g\left(p + \frac{\sqrt{p(1-p)}u}{\sqrt{n}}\right) du. \qquad (2.35)$$

Using elementary algebra and setting $\sigma := \sqrt{p(1-p)}$, it is easy to see that

$$f\left(p + \frac{\sigma u}{\sqrt{n}}\right)^n g\left(p + \frac{\sigma u}{\sqrt{n}}\right) =$$

$$= \frac{p^{np+\sigma u\sqrt{n}}(1-p)^{n(1-p)-\sigma u/\sqrt{n}}\sqrt{p + \dfrac{\sigma u}{\sqrt{n}}}^{\,-1}\sqrt{1 - p - \dfrac{\sigma u}{\sqrt{n}}}^{\,-1}}{\left(p + \dfrac{\sigma u}{\sqrt{n}}\right)^{np+\sigma u\sqrt{n}}\left(1 - p - \dfrac{\sigma u}{\sqrt{n}}\right)^{n(1-p)-\sigma u\sqrt{n}}}$$

$$= \frac{\sqrt{p + \dfrac{\sigma u}{\sqrt{n}}}^{\,-1}\sqrt{1 - p - \dfrac{\sigma u}{\sqrt{n}}}^{\,-1}}{\left(1 + \dfrac{\sigma u}{p\sqrt{n}}\right)^{np+\sigma u\sqrt{n}}\left(1 - \dfrac{\sigma u}{(1-p)\sqrt{n}}\right)^{n(1-p)-\sigma u\sqrt{n}}}$$

$$= \frac{\left(1 + \dfrac{\sigma u}{p\sqrt{n}}\right)^{-pn} \left(1 + \dfrac{\sigma u}{p\sqrt{n}}\right)^{-\sigma u \sqrt{n}} \sqrt{p + \dfrac{\sigma u}{\sqrt{n}}}^{\,-1} \sqrt{1 - p - \dfrac{\sigma u}{\sqrt{n}}}^{\,-1}}{\left(1 - \dfrac{\sigma u}{(1-p)\sqrt{n}}\right)^{(1-p)n} \left(1 - \dfrac{\sigma u}{(1-p)\sqrt{n}}\right)^{-\sigma u \sqrt{n}}}.$$

Since

$$\lim_{n \to \infty} \sqrt{p + \frac{\sigma u}{\sqrt{n}}}^{\,-1} \sqrt{1 - p - \frac{\sigma u}{\sqrt{n}}}^{\,-1} = \frac{1}{\sqrt{p(1-p)}} = \frac{1}{\sigma}$$

and since (2.6) implies that

$$\lim_{n \to \infty} \left(1 + \frac{\sigma u}{p\sqrt{n}}\right)^{\sigma u \sqrt{n}} = \lim_{x \to \infty} \left(\left(1 + \frac{\sigma u/p}{x}\right)^x\right)^{\sigma u} = \left(e^{\sigma u/p}\right)^{\sigma u} = e^{\sigma^2 u^2/p}$$

and, similarly, that

$$\lim_{n \to \infty} \left(1 - \frac{\sigma u}{(1-p)\sqrt{n}}\right)^{-\sigma u \sqrt{n}} = e^{\sigma^2 u^2/(1-p)},$$

it follows that

$$f\left(p + \frac{\sigma u}{\sqrt{n}}\right)^n g\left(p + \frac{\sigma u}{\sqrt{n}}\right) \approx \frac{e^{-u^2}/\sigma}{\left(1 + \dfrac{\sigma u}{p\sqrt{n}}\right)^{pn} \left(1 - \dfrac{\sigma u}{(1-p)\sqrt{n}}\right)^{(1-p)n}}$$

for all sufficiently large values of $n$. Furthermore, as we now set

$$L := \lim_{n \to \infty} \left(1 + \frac{\sigma u}{p\sqrt{n}}\right)^{pn} \left(1 - \frac{\sigma u}{(1-p)\sqrt{n}}\right)^{(1-p)n}$$

$$= \lim_{x \to \infty} \left(1 + \frac{\sigma u}{px}\right)^{px^2} \left(1 - \frac{\sigma u}{(1-p)x}\right)^{(1-p)x^2},$$

we may apply L'Hôpital's rule to infer that

$$\ln(L) = \lim_{x \to \infty} \left(px^2 \ln\left(1 + \frac{\sigma u}{px}\right) + (1-p)x^2 \ln\left(1 - \frac{\sigma u}{(1-p)x}\right)\right)$$

$$= \lim_{x \to \infty} \frac{p \ln\left(1 + \dfrac{\sigma u}{px}\right) + (1-p) \ln\left(1 - \dfrac{\sigma u}{(1-p)x}\right)}{1/x^2}$$

$$= \lim_{x \to \infty} \frac{\dfrac{-\sigma u/x^2}{1 + \dfrac{\sigma u}{px}} + \dfrac{\sigma u/x^2}{1 - \dfrac{\sigma u}{(1-p)x}}}{-2/x^3}$$

$$= \frac{\sigma u}{2} \lim_{x \to \infty} \left( \frac{px^2}{px + \sigma u} - \frac{(1-p)x^2}{(1-p)x - \sigma u} \right)$$

$$= \frac{\sigma u}{2} \lim_{x \to \infty} \frac{-\sigma u x^2}{(px + \sigma u)((1-p)x - \sigma u)}$$

$$= -\frac{\sigma^2 u^2}{2p(1-p)} = -\frac{u^2}{2}.$$

Hence

$$L = e^{-u^2/2}$$

and, by implication,

$$\sum_{\substack{k=0 \\ a_{n,p} \le k \le b_{n,p}}}^{n} \binom{n}{k} p^k (1-p)^{n-k} \approx \frac{\sigma}{\sqrt{2\pi}} \int_a^b f\left(p + \frac{\sigma u}{\sqrt{n}}\right)^n g\left(p + \frac{\sigma u}{\sqrt{n}}\right) du$$

$$\approx \int_a^b \frac{e^{-u^2/2}}{\sqrt{2\pi}} \, du.$$

Note: in a fully rigorous proof we would establish that this concluding approximate equation for large values of $n$ becomes an exact equality as we take the limit as $n$ tends to infinity, and thus we would be able to infer that equation (2.34) is valid, as desired. But even absent this higher level of rigor, our derivation still shows very clearly and compellingly how one of the most amazing facts in all of mathematics—the Central Limit Theorem—can be brought within the reach of standard calculus techniques.

## Exercises

**2.7.3.** Apply Theorem 2.7.1 to find an estimate for the probability that after 1800 tosses of a biased coin with $p = 1/3$ we have a count of at least 590 but no more than 620 heads.

**2.7.4.** Let $X$ be a binomial random variable with parameters $n = 10,000$ and $p = 0.2$.

   a) Use a standard normal approximation to estimate $P(1940 \le X \le 1980)$ (Note: it is required that you interpret the sum representing this probability as a midpoint sum so as to improve the accuracy of the estimate.)

   b) Find the exact value of the probability in a) by means of a binomial-density sum.

## 2.8  Joint Distributions

In the light of what we said in the introductory paragraphs of the preceding section about the universal applicability of the normal density to sums and averages of random values that are produced independently, it is natural and appropriate for us to pose the question as to what exactly it means for random trials to be independent in a concise mathematical sense. As it turns out, the way to go about it is to associate one random variable $X_k : S \to \mathbb{R}$ with each random trial and then to require that for all suitable subsets $B_k \subset \mathbb{R}$ (such as intervals or unions of intervals) the events $E_k = \{s \in S \mid X_k(s) \in B_k\}$ are independent in the sense of Definition 1.5.3. However, in order for us to be able to handle this approach with proper computational efficiency, we need to relate it to the notion a of probability density function, and this in turn requires that we discuss in this present section the notion of a *joint density* of finitely many random variables $X_1, \ldots, X_n$.

**2.8.1 Definition.** If $X_1, \ldots, X_n : S \to \mathbb{R}$ are random variables that are defined on a common sample space $S$, then the *joint cumulative density* (or *joint cumulative distribution function*) of these random variables is the function $F : \mathbb{R}^n \to \mathbb{R}$ that is defined by the equation

$$F(x_1, \ldots, x_n) := P(X_1 \le x_1 \wedge \cdots \wedge X_n \le x_n)$$

$$:= P\left( \bigcap_{k=1}^{n} \{s \in S \mid X_k(s) \le x_k\} \right).$$

**2.8.2 Example.** Assume that a value $s$ is drawn at random from the sample space $S = [0, 1]$ and that all values in $S$ are equally likely to be drawn (that is, the density on $S$ is the uniform density). Given this assumption, we wish to determine the joint cumulative density of the random variables $X, Y : S \to \mathbb{R}$ that are defined by the equations $X(s) := s$ and $Y(s) := 1 - s$, respectively. Using Definition 2.8.1, it follows that

$$F(x, y) = P(X \le x \wedge Y \le y) = P(\{s \in S \mid s \le x\} \cap \{s \in S \mid 1 - s \le y\})$$
$$= P(\{s \in S \mid s \le x\} \cap \{s \in S \mid 1 - y \le s\})$$

for all $(x, y) \in \mathbb{R}^2$, and in order to determine this latter probability, we further observe that $R(X) = R(Y) = [0, 1]$,

$$\{s \in S \mid s \le x\} = \begin{cases} \emptyset & \text{if } x < 0, \\ [0, x] & \text{if } 0 \le x \le 1, \\ [0, 1] & \text{if } x > 1, \end{cases}$$

and

$$\{s \in S \mid 1 - y \le s\} = \begin{cases} \emptyset & \text{if } y < 0, \\ [1 - y, 1] & \text{if } 0 \le y \le 1, \\ [0, 1] & \text{if } y > 1. \end{cases}$$

Hence

$$\{s \in S \mid s \leq x\} \cap \{s \in S \mid 1 - y \leq s\} =$$

$$= \begin{cases} \emptyset & \text{if } x < 0 \text{ or } y < 0, \\ \emptyset & \text{if } x, y \in [0, 1] \text{ and } x < 1 - y, \\ [1 - y, x] & \text{if } x, y \in [0, 1] \text{ and } x \geq 1 - y, \\ [0, x] & \text{if } y > 1 \text{ and } x \in [0, 1], \\ [1 - y, 1] & \text{if } x > 1 \text{ and } y \in [0, 1], \\ [0, 1] & \text{if } x, y > 1, \end{cases}$$

and therefore,

$$F(x, y) = \begin{cases} 0 & \text{if } x < 0 \text{ or } y < 0, \\ 0 & \text{if } x, y \in [0, 1] \text{ and } x < 1 - y, \\ x - (1 - y) = x + y - 1 & \text{if } x, y \in [0, 1] \text{ and } x \geq 1 - y, \\ x & \text{if } y > 1 \text{ and } 0 \leq x \leq 1, \\ 1 - (1 - y) = y & \text{if } x > 1 \text{ and } y \in [0, 1], \\ 1 & \text{if } x, y > 1. \end{cases}$$

A visual representation of this result is shown below in Figure 2.10.

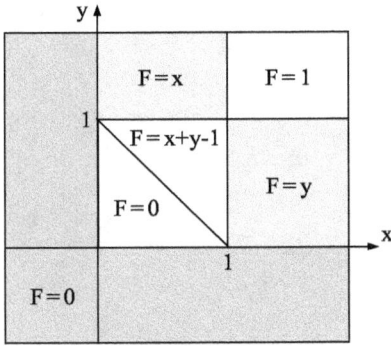

Figure 2.10: domains of definition of the cumulative density $F$.

**2.8.3 Example.** Assume that a value $s$ is drawn at random from the sample space $S = [0, 2]$ and that all values in $S$ are equally likely to be drawn (that is, the density on $S$ is the uniform density). Given this assumption, we wish to determine the joint cumulative density of the random variables $X, Y : S \to \mathbb{R}$ that are defined by the equations $X(s) := s^2$ and $Y(s) := 2 - s$, respectively.

Using Definition 2.8.1, it follows that

$$F(x, y) = P(X \leq x \wedge Y \leq y) = P(\{s \in S \mid s^2 \leq x\} \cap \{s \in S \mid 2 - s \leq y\})$$
$$= P(\{s \in S \mid s^2 \leq x\} \cap \{s \in S \mid 2 - y \leq s\})$$

for all $(x, y) \in \mathbb{R}^2$. To compute this probability, we notice that

$$R(X) = [0, 4],$$
$$R(Y) = [0, 2],$$

and

$$\{s \in S \mid s^2 \leq x\} = \begin{cases} \emptyset & \text{if } x < 0, \\ [0, \sqrt{x}] & \text{if } 0 \leq x \leq 4, \\ [0, 2] & \text{if } x > 4, \end{cases}$$

and

$$\{s \in S \mid 2 - y \leq s\} = \begin{cases} \emptyset & \text{if } y < 0, \\ [2 - y, 2] & \text{if } 0 \leq y \leq 2, \\ [0, 2] & \text{if } y > 2. \end{cases}$$

Hence

$$\{s \in S \mid s^2 \leq x\} \cap \{s \in S \mid 2 - y \leq s\} =$$
$$= \begin{cases} \emptyset & \text{if } x < 0 \text{ or } y < 0, \\ \emptyset & \text{if } x \in [0, 4], y \in [0, 2], \text{ and } \sqrt{x} < 2 - y, \\ [2 - y, \sqrt{x}] & \text{if } x \in [0, 4], y \in [0, 2], \text{ and } \sqrt{x} \geq 2 - y, \\ [0, \sqrt{x}] & \text{if } y > 2 \text{ and } x \in [0, 4], \\ [2 - y, 2] & \text{if } x > 4 \text{ and } y \in [0, 2], \\ [0, 2] & \text{if } x > 4 \text{ and } y > 2, \end{cases}$$

and therefore,

$$F(x, y) = \begin{cases} 0 & \text{if } x < 0 \text{ or } y < 0, \\ 0 & \text{if } x \in [0, 4], y \in [0, 2], \text{ and } \sqrt{x} < 2 - y, \\ (\sqrt{x} - 2 + y)/2 & \text{if } x \in [0, 4], y \in [0, 2], \text{ and } \sqrt{x} \geq 2 - y, \\ \sqrt{x}/2 & \text{if } y > 2 \text{ and } x \in [0, 4], \\ (2 - (2 - y))/2 = y/2 & \text{if } x > 4 \text{ and } y \in [0, 2], \\ 1 & \text{if } x > 4 \text{ and } y > 2. \end{cases}$$

A visual representation of this result is shown below in Figure 2.11.

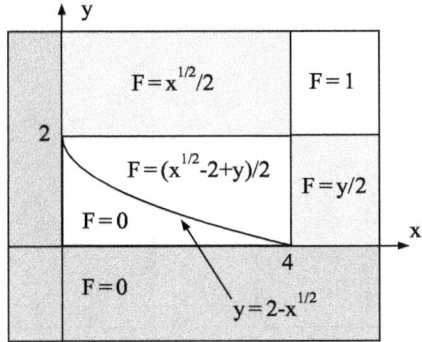

Figure 2.11: domains of definition of the cumulative density $F$.

**2.8.4 Example.** Assume that two values $s$ and $t$ are chosen at random from the interval $(0, 1]$ and that all values in $(0, 1]$ are equally likely to be picked. Then the sample space for the pair of random values $(s, t)$ is $S = (0, 1] \times (0, 1]$, and the equations $X(s, t) := s$ and $Y(s, t) := st$ define two random variables $X, Y : S \rightarrow \mathbb{R}$. Given this setup, the cumulative density of $X$ and $Y$ satisfies the equation

$$F(x, y) = P(X \leq x \wedge Y \leq y) = P(\{(s, t) \in S \mid s \leq x\} \cap \{(s, t) \in S \mid st \leq y\})$$
$$= P(\{(s, t) \in S \mid s \leq x\} \cap \{(s, t) \in S \mid t \leq y/s\})$$

for all $(x, y) \in \mathbb{R}^2$. Furthermore, since the range of both $X$ and $Y$ is evidently $(0, 1]$, the pictures of the events

$$E_x := \{(s, t) \in S \mid s \leq x\}$$

and

$$F_y := \{(s, t) \in S \mid t \leq y/s\}$$

in Figure 2.12, in the case where both $x$ and $y$ are in $(0, 1] = R(X) = R(Y)$, allow us to infer that

$$F(x, y) = \begin{cases} P(\emptyset) = 0 & \text{if } x \leq 0 \text{ or } y \leq 0, \\ y + \int_y^x (y/s) \, ds = y + y \ln(x/y) & \text{if } 0 < y \leq x \leq 1, \\ P(E_x) = x & \text{if } 0 < x \leq y \leq 1, \\ P(E_x) = x & \text{if } 0 < x \leq 1 \text{ and } y > 1, \\ P(F_y) = y + \int_y^1 (y/s) \, ds = y - y \ln(y) & \text{if } x > 1 \text{ and } 0 < y \leq 1, \\ P(S) = 1 & \text{if } x, y > 1. \end{cases}$$

for all $(x, y) \in \mathbb{R}^2$ (see Figure 2.13 for a visual representation).

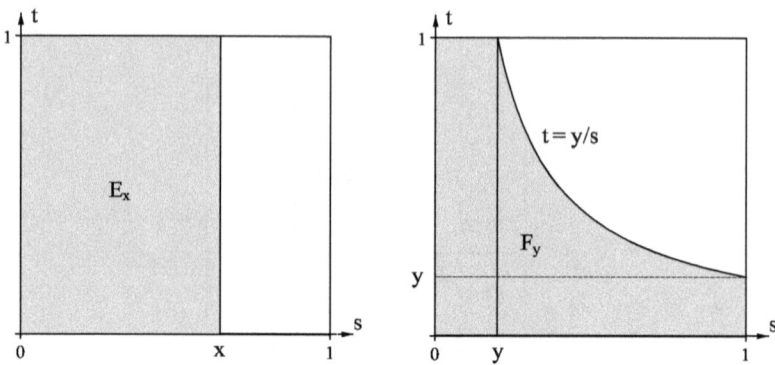

Figure 2.12: the events $E_x$ and $F_y$.

**2.8.5 Example.** Assume that $G$ is the cumulative density of a random variable $X : S \to \mathbb{R}$, that is, $G(x) = P(X \leq x)$ for all $x \in \mathbb{R}$. Setting $Y := X$, we wish to find the joint cumulative density of $X$ and $Y$. Using Definition 2.8.1, it follows that

$$F(x, y) = P(X \leq x \wedge Y \leq y) = P(X \leq x \wedge X \leq y)$$

for all $x, y \in \mathbb{R}$. Furthermore, if $X$ is less than or equal to $x$ and also to $y$, then $X$ is less than or equal to the smaller of these two values. Consequently, the statement $X \leq x \wedge X \leq y$ is equivalent to $X \leq \min\{x, y\}$, and therefore,

$$F(x, y) = P(X \leq \min\{x, y\}) = G(\min\{x, y\}),$$

or equivalently,

$$F(x, y) = \begin{cases} G(x) & \text{if } x \leq y, \\ G(y) & \text{if } y < x. \end{cases}$$

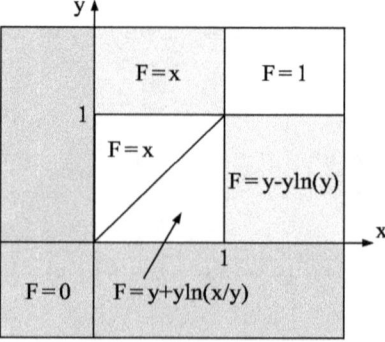

Figure 2.13: domains of definition of the cumulative density $F$.

Having introduced the notion of a cumulative joint density, it is perfectly natural for us to agree that a function $f : \mathbb{R}^n \to \mathbb{R}$ is a joint density of *continuous* random variables $X_1, \ldots, X_n : S \to \mathbb{R}$ if—in generalization of (2.10)—the cumulative density $F(x_1, \ldots, x_n) = P(X_1 \leq x_1 \wedge \cdots \wedge X_n \leq x_n)$ satisfies the equation

$$F(x_1, \ldots, x_n) = \int_{-\infty}^{x_1} \cdots \int_{-\infty}^{x_n} f(t_1, \ldots, t_n) \, dt_n \ldots dt_1 \qquad (2.36)$$

for all $(x_1, \ldots, x_n) \in \mathbb{R}^n$. Given this defining equation for $f$, it follows that

$$1 = \int_{-\infty}^{\infty} \cdots \int_{-\infty}^{\infty} f(x_1, \ldots, x_n) \, dx_n \ldots dx_1 \qquad (2.37)$$

because clearly, $P(X_1 \leq \infty \wedge \cdots \wedge X_n \leq \infty) = 1$. Furthermore, if the random variables $X_1, \ldots, X_n : S \to \mathbb{R}$ are discrete (that is, if the ranges $R(X_1), \ldots, R(X_n)$ can each be listed), then, in direct analogy to Definition 2.2.1, we set

$$E_{x_1, \ldots, x_n} = \{ s \in S \mid X_1(s) = x_1 \wedge \cdots \wedge X_n = x_n \} = \bigcap_{k=1}^{n} \{ s \in S \mid X_k(s) = x_k \},$$

for all $(x_1, \ldots, x_n) \in \mathbb{R}^n$ and define the discrete *joint density function* $f : \mathbb{R}^n \to \mathbb{R}$ of the random variables $X_1, \ldots, X_n$ via the equation

$$f(x_1, \ldots, x_n) := P(X_1 = x_1 \wedge \cdots \wedge X_n = x_n) := P(E_{x_1, \ldots, x_n}).$$

**2.8.6 Example.** Assume that a value $s$ is chosen at random from the sample space $S = [0, 2]$ and that all values in $S$ are equally likely to be picked. Then the joint density of the discrete random variables

$$X(s) := \begin{cases} 1 & \text{if } s \in [0, 1), \\ 3 & \text{if } s \in [1, 2], \end{cases}$$

and

$$Y(s) := \begin{cases} 0 & \text{if } s \in [0, 2/3), \\ 2 & \text{if } s \in [2/3, 3/2), \\ 1 & \text{if } s \in [3/2, 2], \end{cases}$$

is

$$f(x,y) = \begin{cases} P([0,1) \cap [0,2/3)) = P([0,2/3)) = 1/3 & \text{if } x = 1 \text{ and } y = 0, \\ P([0,1) \cap [2/3,3/2)) = P([2/3,1)) = 1/6 & \text{if } x = 1 \text{ and } y = 2, \\ P([0,1) \cap [3/2,2)) = P(\emptyset) = 0 & \text{if } x = 1 \text{ and } y = 1, \\ P([1,2] \cap [0,2/3)) = P(\emptyset) = 0 & \text{if } x = 3 \text{ and } y = 0, \\ P([1,2] \cap [2/3,3/2)) = P([1,3/2)) = 1/4 & \text{if } x = 3 \text{ and } y = 2, \\ P([1,2] \cap [3/2,2)) = P([3/2,2]) = 1/4 & \text{if } x = 3 \text{ and } y = 1, \\ 0 & \text{otherwise.} \end{cases}$$

**2.8.7 Theorem.** *If $f$ is a joint density of finitely many continuous random variables $X_1, \ldots, X_n : S \to \mathbb{R}$, then*

$$P(a_1 \leq X_1 \leq b_1 \wedge \cdots \wedge a_n \leq X_n \leq b_n) = \int_{a_1}^{b_1} \cdots \int_{a_n}^{b_n} f(x_1, \ldots, x_n) \, dx_n \ldots dx_1$$

*for all $a_1, \ldots, a_n, b_1, \ldots, b_n \in \mathbb{R}$ with $a_k \leq b_k$ for all $k \in \{1, \ldots, n\}$. Furthermore, if $f$ is continuous (in the standard calculus sense) and if $F$ is the joint cumulative density of $X_1, \ldots, X_n$, then*

$$f(x_1, \ldots, x_n) = \frac{\partial^n}{\partial x_n \ldots \partial x_1} F(x_1, \ldots, x_n).$$

*Proof.* We will consider only the case $n = 2$ as the proof for arbitrary values of $n$ is largely analogous. So we will show that

$$P(a \leq X \leq b \wedge c \leq Y \leq d) = \int_a^b \int_c^d f(x,y) \, dy \, dx$$

for all $a, b, c, d \in \mathbb{R}$ with $a \leq b$ and $c \leq d$. To do so, we set

$$E_{a,c} := \{s \in S \mid X(s) \leq a \wedge Y(s) \leq c\},$$
$$E_{b,d} := \{s \in S \mid X(s) \leq b \wedge Y(s) \leq d\},$$
$$E_{a,d} := \{s \in S \mid X(s) \leq a \wedge Y(s) \leq d\},$$
$$E_{b,c} := \{s \in S \mid X(s) \leq b \wedge Y(s) \leq c\},$$

and observe that

$$F := \{s \in S \mid a \leq X(s) \leq b \wedge c \leq Y(s) \leq d\} = E_{b,d} \setminus (E_{a,d} \cup E_{b,c}).$$

Since $E_{a,d} \cup E_{b,c} \subset E_{b,d}$, it follows that $E_{b,d}$ is the disjoint union of $E_{a,d} \cup E_{b,c}$ and $F = E_{b,d} \setminus (E_{a,d} \cup E_{b,c})$, and therefore,

$$\begin{aligned} P(E_{b,d}) &= P(E_{a,d} \cup E_{b,c}) + P(F) \\ &= P(E_{a,d}) + P(E_{b,c}) - P(E_{a,d} \cap E_{b,c}) + P(F) \\ &= P(E_{a,d}) + P(E_{b,c}) - P(E_{a,c}) + P(F). \end{aligned}$$

Using (2.36), we may infer that

$$P(F) = P(E_{b,d}) - P(E_{a,d}) - P(E_{b,c}) + P(E_{a,c})$$

$$= \int_{-\infty}^{b} \int_{-\infty}^{d} f(x,y)\, dy\, dx - \int_{-\infty}^{a} \int_{-\infty}^{d} f(x,y)\, dy\, dx$$

$$- \int_{-\infty}^{b} \int_{-\infty}^{c} f(x,y)\, dy\, dx + \int_{-\infty}^{a} \int_{-\infty}^{c} f(x,y)\, dy\, dx$$

$$= \int_{a}^{b} \int_{-\infty}^{d} f(x,y)\, dy\, dx - \int_{a}^{b} \int_{-\infty}^{c} f(x,y)\, dy\, dx$$

$$= \int_{a}^{b} \int_{c}^{d} f(x,y)\, dy\, dx,$$

as desired. Furthermore, using (2.36) in conjunction with the fundamental theorem of calculus, we find that

$$\frac{\partial^2}{\partial y \partial x} F(x,y) = \frac{\partial^2}{\partial y \partial x} \int_{-\infty}^{x} \int_{-\infty}^{y} f(u,v)\, dv\, du = \frac{\partial}{\partial y} \int_{-\infty}^{y} f(x,v)\, dv = f(x,y)$$

whenever $f$ is continuous. □

**2.8.8 Corollary.** *If $f : \mathbb{R}^n \to \mathbb{R}$ is the joint density of given random variables $X_1, \ldots, X_n : S \to \mathbb{R}$ and if $f$ is continuous, then $f(x_1, \ldots, x_n) \geq 0$ for all $(x_1, \ldots, x_n) \in \mathbb{R}^n$.*

*Proof.* If there did exist a point $(x_1, \ldots, x_n) \in \mathbb{R}^n$ at which $f$ is negative, then, by continuity, there would exist a (small) number $\varepsilon > 0$ such that $f(u_1, \ldots, u_n) < 0$ for all $(u_1, \ldots, u_n) \in R_\varepsilon := [x_1, x_1 + \varepsilon] \times \cdots \times [x_n, x_n + \varepsilon]$. Hence

$$P(x_1 \leq X_1 \leq x_1 + \varepsilon \wedge \cdots \wedge x_n \leq X_n \leq x_n + \varepsilon)$$

$$= \int_{R_\varepsilon} f(u_1, \ldots, u_n)\, du_n \ldots du_1 < 0.$$

But this is impossible because no probability can ever be negative. Thus there cannot exist a point in $\mathbb{R}^n$ at which $f$ is negative. □

*Remark.* Since probabilities are always non-negative, probability densities are naturally non-negative as well. But strictly speaking, this property of non-negativity can be rigorously inferred only if $f$ is assumed to be continuous. That is to say, for discontinuous densities it has to be imposed by definition.

*Remark.* In direct analogy to (2.15), the meaning of a joint density function of continuous random variables $X_1, \ldots, X_n$ can also be captured, somewhat informally, by the equation

$$P(X_1 \in [x_1, x_1 + dx_1] \wedge \cdots \wedge X_n \in [x_n, x_n + dx_n])$$
$$= f(x_1, \ldots, x_n) \, dx_1 \ldots dx_n.$$

To justify this equation in the case where $n = 2$, we may argue as follows: since

$$[x, x + \Delta x] \times [y, y + \Delta y] = (-\infty, x + \Delta x] \times (-\infty, y + \Delta y] \smallsetminus$$
$$\smallsetminus \, ((-\infty, x) \times (-\infty, y + \Delta y] \cup (-\infty, x + \Delta x] \times (-\infty, y))$$

for all $x, y \in \mathbb{R}$ and $\Delta x, \Delta y > 0$, we may infer that

$$\frac{P(X \in [x, x + dx] \wedge Y \in [y, y + dy])}{dx \, dy} = \frac{P((X, Y) \in [x, x + dx] \times [y, y + dy])}{dx \, dy}$$

$$= \lim_{\Delta x, \Delta y \to 0^+} \frac{P((X, Y) \in [x, x + \Delta x] \times [y, y + \Delta y])}{\Delta x \Delta y}$$

$$= \lim_{\Delta x, \Delta y \to 0^+} \frac{F(x + \Delta x, y + \Delta y) - F(x + \Delta x, y) - F(x, y + \Delta y) + F(x, y)}{\Delta x \Delta y}$$

$$= \lim_{\Delta x, \Delta y \to 0^+} \frac{\dfrac{F(x + \Delta x, y + \Delta y) - F(x + \Delta x, y)}{\Delta y} - \dfrac{F(x, y + \Delta y) - F(x, y)}{\Delta y}}{\Delta x}$$

$$= \lim_{\Delta x \to 0^+} \frac{\partial F / \partial y (x + \Delta x, y) - \partial F / \partial y (x, y)}{\Delta x} = \frac{\partial^2 F}{\partial x \partial y}(x, y) = f(x, y),$$

and therefore,

$$P(X \in [x, x + dx] \wedge Y \in [y, y + dy]) = f(x, y) \, dx \, dy, \qquad (2.38)$$

as desired.

*Remark.* Equation (2.38) implies that for any (sufficiently regular) region $R \subset \mathbb{R}^2$ it is the case that

$$\boxed{P((X, Y) \in R) = \int_R f(x, y) \, dx \, dy,} \qquad (2.39)$$

because any such region $R$ can be construed to be a disjoint union of infinitely many infinitesimal rectangles $[x, x + dx) \times [y, y + dy)$ and because the total probability for $(X, Y)$ to be contained in any of these rectangles is naturally given by the integral in (2.39). Similarly and more generally, it also is the case that

$$\boxed{P((X_1, \ldots, X_n) \in R) = \int_R f(x_1, \ldots, x_n) \, dx_1 \ldots dx_n}$$

for any (sufficiently regular) region $R \subset \mathbb{R}^n$.

**2.8.9 Example.** In Example 2.8.2 we found the joint cumulative density of the random variables $X(s) = s$ and $Y(s) = 1 - s$ to be

$$F(x,y) = \begin{cases} 0 & \text{if } x < 0 \text{ or } y < 0, \\ 0 & \text{if } 0 \le x, y \le 1 \text{ and } x < 1 - y, \\ x - (1 - y) = x + y - 1 & \text{if } 0 \le x, y \le 1 \text{ and } x \ge 1 - y, \\ x & \text{if } y > 1 \text{ and } 0 \le x \le 1, \\ 1 - (1 - y) = y & \text{if } x > 1 \text{ and } 0 \le y \le 1, \\ 1 & \text{if } x, y > 1. \end{cases}$$

However, as we take the mixed partial derivative of this cumulative density, we readily find that

$$f(x,y) = \frac{\partial^2}{\partial x \partial y} F(x,y) = 0$$

for all $(x, y) \in \mathbb{R}^2$ at which $F$ is twice differentiable. Consequently, according to Theorem 2.8.7, we arrive at the nonsensical conclusion that all probabilities of the form $P(a \le X \le b \wedge c \le Y \le d)$ are equal to zero. So what this example illustrates is the general fact that joint densities in the spirit of Theorem 2.8.7 do not always exist. There always is a joint *cumulative* density, but a joint regular density that satisfies the defining equation (2.36) and the statement of Theorem 2.8.7 thereby may be impossible to find.

*Remark.* As we examine the preceding example a little more closely, we come to realize quite readily that there does exist a joint density function of $X$ and $Y$ that is non-zero (or supported), not on an extended two-dimensional region in $\mathbb{R}^2$, but rather on the one-dimensional straight line segment $L$ that connects the points $(1, 0)$ and $(0, 1)$ and is described by the equation $x + y = 1$. More precisely, the probability $P(a \le X \le b \wedge c \le Y \le d)$ turns out to be equal to the length of the line segment $[a, b] \times [c, d] \cap L$ (which is zero if the intersection is empty) divided by the length of $L$ (which is $\sqrt{2}$). In other words and by implication, the joint density $f$ turns out to be given by the equation

$$f(x,y) = \begin{cases} 1/\sqrt{2} & \text{if } x + y = 1 \text{ and } x \in [0, 1], \\ 0 & \text{otherwise} \end{cases}$$

for all $(x, y) \in \mathbb{R}^2$. However, a more in-depth discussion of the meaning of this defining equation and its potential generalizations to probability densities that are supported on lower-dimensional subsets of $\mathbb{R}^n$ would lead us too far and will therefore be omitted.

**2.8.10 Example.** Given the joint cumulative density

$$F(x,y) = \begin{cases} 0 & \text{if } x \le 0 \text{ or } y \le 0, \\ y + y\ln(x/y) & \text{if } 0 < y \le x \le 1, \\ x & \text{if } 0 < x \le y \le 1, \\ x & \text{if } 0 < x \le 1 \text{ and } y > 1, \\ y - y\ln(y) & \text{if } x > 1 \text{ and } 0 < y \le 1, \\ 1 & \text{if } x, y > 1. \end{cases}$$

that we computed in Example 2.8.4, we find that the corresponding density is

$$f(x,y) = \frac{\partial^2}{\partial x \partial y} F(x,y) = \begin{cases} 1/x & \text{if } 0 < y \le x \le 1, \\ 0 & \text{otherwise.} \end{cases}$$

**2.8.11 Example.** Assume that a point $(s,t)$ is chosen at random from the rectangle $S = [a,b] \times [c,d]$ and that all points are equally likely to be picked. In this case we naturally expect the joint density of the random variables $X(s,t) := s$ and $Y(s,t) := t$ to be the uniform density on $S$, and in order to confirm this expectation, we set

$$A := (b-a)(d-c)$$

and observe that the joint cumulative density of $X$ and $Y$ is

$$F(x,y) = \begin{cases} P(\emptyset) = 0 & \text{if } x < a \text{ or } y < c, \\ P([a,x] \times [c,y]) = (x-a)(y-c)/A & \text{if } x \in [a,b] \text{ and } y \in [c,d], \\ P([a,x] \times [c,d]) = (x-a)(d-c)/A & \text{if } x \in [a,b] \text{ and } y > d, \\ P([a,b] \times [c,y]) = (b-a)(y-c)/A & \text{if } x > b \text{ and } y \in [c,d], \\ P(S) = 1 & \text{if } x > b \text{ and } y < d. \end{cases}$$

This yields

$$f(x,y) = \frac{\partial^2}{\partial x \partial y} F(x,y) = \begin{cases} 1/A & \text{if } x \in [a,b] \text{ and } y \in [c,d], \text{ i.e., } (x,y) \in S, \\ 0 & \text{otherwise,} \end{cases}$$

as conjectured.

As we will see in the next section, two random variables $X$ and $Y$ with joint density $f$ are independent (see the introduction to this section) if $f$ is equal to the product of the individual densities of $X$ and $Y$. Hence we introduce the following definition:

**2.8.12 Definition.** If $f : \mathbb{R}^2 \to \mathbb{R}$ is the joint density of two continuous random variables $X, Y : S \to \mathbb{R}$, then the corresponding marginal densities of $X$ and $Y$ are defined by the equations

$$f_X(x) := \int_{-\infty}^{\infty} f(x, y)\, dy,$$

$$f_Y(y) := \int_{-\infty}^{\infty} f(x, y)\, dx$$

for all $(x, y) \in \mathbb{R}^2$. More generally, if $f$ is the joint density of given continuous random variables $X_1, \ldots, X_n : S \to \mathbb{R}$, then, for any $k \in \{1, \ldots, n\}$, the marginal density of $X_k$ at any point $x_k \in \mathbb{R}$ is

$$f_{X_k}(x_k) := \int_{\mathbb{R}^{n-1}} f(u_1, \ldots, u_{k-1}, x_k, u_{k+1}, \ldots, u_n)\, du_1 \ldots du_{k-1}\, du_{k+1} \ldots du_n.$$

Furthermore, the corresponding discrete-case formulas are

$$f_X(x) := \sum_{y \in R(Y)} f(x, y),$$

$$f_Y(y) := \sum_{x \in R(X)} f(x, y),$$

and

$$f_{X_k}(x_k) := \sum_{(u_1, \ldots, u_{k-1}, u_{k+1}, \ldots, u_n) \in R} f(u_1, \ldots, u_{k-1}, x_k, u_{k+1}, \ldots, u_n).$$

where

$$R := R(X_1) \times \cdots \times R(X_{k-1}) \times R(X_{k+1}) \times \cdots \times R(X_n).$$

*Remark.* If $f$ is the joint density of two continuous random variables $X, Y : S \to \mathbb{R}$, then the cumulative density of $X$ is

$$F(x) = P(X \le x) = P(X \le x \wedge Y \in \mathbb{R}) = \int_{-\infty}^{x} \int_{-\infty}^{\infty} f(u, v)\, dv\, du$$

$$= \int_{-\infty}^{x} f_X(u)\, du,$$

and therefore, the density of $X$ is

$$F'(x) = \frac{d}{dx} \int_{-\infty}^{x} f_X(u)\, du = f_X(x).$$

In other words, the marginal density of $X$ is identical with the density of $X$ as given by the defining equations (2.10) and (2.11). Similarly it can be shown that the marginal density $f_{X_K}$ that is associated with the joint density $f$ of finitely many continuous random variables $X_1, \ldots, X_n : S \to \mathbb{R}$ is the density of $X_k$. Moreover, the same conclusion is easily seen to be valid in the discrete case as well.

**2.8.13 Example.** If

$$f(x, y) = \frac{e^{-((x-\mu)^2 + (y-\nu)^2)/(2\sigma^2)}}{2\pi\sigma^2},$$

happens to be the joint normal density of two random variables $X, Y : S \to \mathbb{R}$, then

$$f_X(x) = \int_{-\infty}^{\infty} \frac{e^{-((x-\mu)^2 + (y-\nu)^2)/(2\sigma^2)}}{2\pi\sigma^2} \, dy$$

$$= \frac{e^{-(x-\mu)^2/(2\sigma^2)}}{\sqrt{2\pi}\sigma} \int_{-\infty}^{\infty} \frac{e^{-(y-\nu)^2/(2\sigma^2)}}{\sqrt{2\pi}\sigma} \, dy$$

$$= \frac{e^{-(x-\mu)^2/(2\sigma^2)}}{\sqrt{2\pi}\sigma}$$

because the integral from $-\infty$ to $\infty$ of the normal density function

$$\frac{e^{-(y-\nu)^2/(2\sigma^2)}}{\sqrt{2\pi}\sigma}$$

is equal to one.

**2.8.14 Example.** Given the joint density

$$f(x, y) = \begin{cases} 1/x & \text{if } 0 < y \le x \le 1, \\ 0 & \text{otherwise,} \end{cases}$$

that we found in Example 2.8.10, it follows that the corresponding marginal densities are

$$f_X(x) = \int_{-\infty}^{\infty} f(x, y) \, dy = \begin{cases} \int_0^x 1/x \, dy = 1 & \text{if } x \in (0, 1], \\ 0 & \text{otherwise} \end{cases}$$

and

$$f_Y(y) = \int_{-\infty}^{\infty} f(x, y) \, dx = \begin{cases} \int_y^1 1/x \, dx = -\ln(y) & \text{if } y \in (0, 1], \\ 0 & \text{otherwise.} \end{cases}$$

In accordance with the preceding remark, the latter equation for $f_Y$ is consistent with the result found earlier in Example 2.5.6.

In order to understand how joint density functions can be used to compute expectations, let us assume that $f$ is the joint density of two random variables $X, Y : S \to \mathbb{R}$ and that $H$ is a real-valued function defined on $\mathbb{R}^2$. Then the function $H(X, Y)$ which, by definition, assigns to every outcome $s \in S$ the output value $H(X(s), Y(s))$ is a real-valued function defined on $S$ and as such it is a random variable. In order to find the expected value of this random variable, we observe that the probability for the point $(X(s), Y(s))$ to be contained in a small rectangle $[x, x + \Delta x] \times [y, y + \Delta y]$ is approximately equal to $f(x, y)\Delta x \Delta y$. Consequently, as we subdivide $\mathbb{R}^2$, in standard calculus fashion, into infinitely many rectangles

$$[x_i, x_{i+1}] \times [y_j, y_{j+1}] = [x_i, x_i + \Delta x] \times [y_j, y_j + \Delta y],$$

we find that the expected value of $H(X, Y)$ is approximately equal to

$$\sum_{i=-\infty}^{\infty} \sum_{j=-\infty}^{\infty} H(x_i, y_j) f(x_i, y_j) \Delta x \Delta y.$$

By implication, if $H$ and $f$ are properly integrable, we may take the limit as $\Delta x$ and $\Delta y$ converge to zero to infer that

$$E(H(X, Y)) = \int_{-\infty}^{\infty} \int_{-\infty}^{\infty} H(x, y) f(x, y)\, dx\, dy. \tag{2.40}$$

Alternatively and in direct appeal to formula (2.17), we may also write (in generic form)

$$E(H(X, Y)) = \int_S H(X(s), Y(s)) p(s)\, ds. \tag{2.41}$$

The natural generalizations of these formulas to finitely many random variables $X_1, \ldots, X_n$ are

$$E(H(X_1, \ldots, X_n)) = \int_{-\infty}^{\infty} \cdots \int_{-\infty}^{\infty} H(x_1, \ldots, x_n) f(x_1, \ldots, x_n)\, dx_1 \ldots dx_n$$

and

$$E(H(X_1, \ldots, X_n)) = \int_S H(X_1(s), \ldots, X_n(s)) p(s)\, ds.$$

**2.8.15 Example.** Given the function $H(x, y) := x^2 y$, we wish to find the expectation of $H(X, Y)$ for the random variables $X(s, t) = s$ and $Y(s, t) = st$ defined in Example 2.8.4. To do so, we apply formula (2.40) to the joint density $f$, as determined in Example 2.8.10. This yields

$$E(H(X, Y)) = \int_{-\infty}^{\infty} \int_{-\infty}^{\infty} x^2 y f(x, y)\, dy\, dx = \int_0^1 \int_0^x \frac{x^2 y}{x}\, dy\, dx$$

$$= \int_0^1 \frac{x y^2}{2}\Big|_0^x\, dx = \int_0^1 \frac{x^3}{2}\, dx = \frac{1}{8}.$$

Alternatively, in applying (2.41), the same result can also be found by means of an integral over the sample space $S = [0,1] \times [0,1]$ if the (generic) infinitesimal probability $p(s)\,ds$ is replaced by $1 \cdot ds\,dt$:

$$E(H(X,Y)) = \int_0^1 \int_0^1 X(s,t)^2 Y(s,t)\,ds\,dt = \int_0^1 \int_0^1 s^3 t\,ds\,dt = \int_0^1 \frac{t}{4}\,dt = \frac{1}{8}.$$

Furthermore, to compute the variance of $H(X,Y)$, we observe that

$$E(H(X,Y)^2) = \int_{-\infty}^{\infty} \int_{-\infty}^{\infty} x^4 y^2 f(x,y)\,dy\,dx = \int_0^1 \int_0^x x^3 y^2\,dy\,dx$$

$$= \int_0^1 \frac{x^3 y^3}{3}\Big|_0^x\,dx = \int_0^1 \frac{x^6}{3}\,dx = \frac{1}{21},$$

or alternatively,

$$E(H(X,Y)^2) = \int_0^1 \int_0^1 X(s,t)^4 Y(s,t)^2\,ds\,dt = \int_0^1 \int_0^1 s^6 t^2\,ds\,dt = \frac{1}{21}.$$

Hence

$$\mathrm{Var}(H(X,Y)) = E(H(X,Y)^2) - E(H(X,Y))^2 = \frac{1}{21} - \frac{1}{64} = \frac{43}{1344}.$$

*Remark.* If $X, Y : S \to \mathbb{R}$ are discrete random variables, then the formula analogous to (2.40) is

$$\boxed{E(H(X,Y)) = \sum_{y \in R(Y)} \sum_{x \in R(X)} H(x,y) f(x,y),} \qquad (2.42)$$

and if $S$ is discrete as well, then (2.17) assumes the form

$$\boxed{E(H(X,Y)) = \sum_{s \in S} H(X(s), Y(s)) P(\{s\}).} \qquad (2.43)$$

## Exercises

**2.8.16.** Let $X(s,t) := 1 - s$ and $Y(s,t) := t^2$ for all $(s,t) \in [0,2] \times [1,4]$. Find the joint density of $X$ and $Y$.

**2.8.17.** Let $X(s) := s^2 + 1$ and $Y(s) := (s-1)^2$ for all $s \in [0,1]$. Find the joint cumulative density of $X$ and $Y$.

**2.8.18.** Find formulas analogous to (2.42) and (2.43) for the expectation of $H(X_1, \ldots, X_n)$ in the case where $X_1, \ldots, X_n : S \to \mathbb{R}$ are discrete random variables (and where $S$ is potentially discrete as well).

**2.8.19.** Find the joint cumulative density of the random variables $X, Y :$ $[0, 1] \to \mathbb{R}$ that are defined by the equations $X(s) = s^2$ and $Y(s) = s$, and explain why in this case a continuous joint density does not exist.

**2.8.20.** Explain why there is no continuous joint density for the random variables $X$ and $Y$ in Example 2.8.3.

**2.8.21.** Find the joint density and the joint cumulative density of the random variables $X, Y : \mathbb{R}^2 \to \mathbb{R}$ that are defined by the equations $X(s, t) := s^2$ and $Y(s, t) := t^2$. You may assume the probability of an event $E$ in the sample space $\mathbb{R}^2$ to be $P(E) = \int_E p(s, t) \, ds \, dt$ where $p(s, t) := e^{-(s^2 + t^2)/2}/(2\pi)$ for all $(s, t) \in \mathbb{R}^2$.

**2.8.22.** Use your answer to Exercise 2.8.21 to determine $E(\sqrt{XY})$.

**2.8.23.** Verify the value for the expectation that you found in Exercise 2.8.22 by means of an integral over the sample space $\mathbb{R}^2$. (Note: in this case the generic expression $p(s)ds$ in the formula $E(X) = \int_S X(s)p(s) \, ds$ assumes the form $p(s, t) \, ds \, dt$.)

**2.8.24.** Find the (discrete) joint density of the random variables $X, Y : [0, 1] \to \mathbb{R}$ that are defined as follows:

$$X(t) := \begin{cases} 1 & \text{if } t \in [0, 1/2], \\ 2 & \text{if } t \in (1/2, 1], \end{cases} \quad \text{and} \quad Y(t) := \begin{cases} 3 & \text{if } t \in [0, 1/3], \\ 1 & \text{if } t \in (1/3, 3/4], \\ 4 & \text{if } t \in (3/4, 1]. \end{cases}$$

**2.8.25.** Assume that for some $n \in \mathbb{N}$ the discrete joint density of the random variables $X, Y : S \to \mathbb{R}$ is

$$f(i, j) := \begin{cases} 2/(n(n+1)) & \text{if } 1 \leq j \leq i \leq n \\ 0 & \text{otherwise.} \end{cases}$$

Find the marginal densities of $X$ and $Y$ and use these marginal densities to find the expected values of $X$ and $Y$.

**2.8.26.** Assume that the joint density of two random variables $X$ and $Y$ is

$$f(x, y) = \begin{cases} Ce^{-x-2y} & \text{for } x, y \geq 0 \\ 0 & \text{otherwise.} \end{cases}$$

Use (2.37) to find the value of $C$ and determine the marginal densities $f_X(x)$ and $f_Y(y)$.

**2.8.27.** Use the marginal densities determined in Exercise 2.8.26 to find the expected value of $X$.

**2.8.28.** Use the density given in Exercise 2.8.26 to find the expected value of $X^2 Y$.

**2.8.29.** Assume that the joint density of two random variables $X$ and $Y$ is

$$f(x, y) = Ce^{-((x-1)^2 + (y+2)^2)/8}$$

for all $x, y \in \mathbb{R}^2$. Find the value of $C$ and determine the marginal densities $f_X(x)$ and $f_Y(y)$.

**2.8.30.** Use integration in polar coordinates in conjunction with (2.39) to determine $P(X^2 + Y^2 - 2X + 4Y \leq -4)$ under the additional assumption that the joint density of $X$ and $Y$ is the density given in Exercise 2.8.29.

## 2.9 Independent Random Variables

As indicated in the introductory paragraph of Section 2.8, the independence of finitely many random variables $X_1, \ldots, X_n : S \to \mathbb{R}$ is to be understood in terms of the independence of events of the form

$$E_k = \{X_k \in B_k\} = \{s \in S \mid X_k(s) \in B_k\}.$$

Unfortunately, in order for this seemingly very natural approach to be feasible in a rigorous and fully consistent manner, the sets $B_k \subset \mathbb{R}$ cannot be chosen completely arbitrarily but must be elements of certain classes of subsets of $\mathbb{R}$ that prominently include intervals and finite (or discretely infinite) unions of intervals. However, since the details of this purely theoretical issue are clearly beyond the scope of the present exposition, we will forego a precise definition and simply refer to these sets $B_k$ as 'suitable'.

**2.9.1 Definition.** We say that random variables $X_1, \ldots, X_n : S \to \mathbb{R}$ are *independent* if the events $\{X_1 \in B_1\}, \ldots, \{X_n \in B_n\}$ are independent for all suitable sets $B_1, \ldots, B_n \subset \mathbb{R}$ (such as intervals or discrete unions of intervals).

**2.9.2 Theorem.** *The random variables* $X_1, \ldots, X_n : S \to \mathbb{R}$ *are independent if and only if*

$$P\left(\bigcap_{i=1}^{n} \{X_i \in B_i\}\right) = \prod_{i=1}^{n} P(X_i \in B_i)$$

*for all suitable subsets* $B_1, \ldots, B_n \subset \mathbb{R}$.

*Proof.* Since independence trivially implies the validity of the equation above, it only remains to be shown that the reverse conclusion is valid as well. To this end we need to prove (see Definition 1.5.3) that for any index set $I \subset \{1, \ldots, n\}$ it is the case that

$$P\left(\bigcap_{i \in I} \{X_i \in B_i\}\right) = \prod_{i \in I} P(X_i \in B_i).$$

Setting $B_j := \mathbb{R}$ for all $j \in \{1, \ldots, n\} \setminus I$, it follows that

$$\{X_j \in B_j\} = S$$

for all $j \in \{1, \ldots, n\} \setminus I$, and therefore,

$$P\left(\bigcap_{i \in I} \{X_i \in B_i\}\right) = P\left(\bigcap_{i=1}^{n} \{X_i \in B_i\}\right) = \prod_{i=1}^{n} P(X_i \in B_i)$$

$$= \prod_{i \in I} P(X_i \in B_i) \prod_{j \in \{1, \ldots, n\} \setminus I} P(S) = \prod_{i \in I} P(X_i \in B_i)$$

because $P(S) = 1$. □

**2.9.3 Theorem.** *If $f : \mathbb{R}^2 \to \mathbb{R}$ is the joint density of two (discrete or continuous) random variables $X, Y : S \to \mathbb{R}$, then $X$ and $Y$ are independent if and only if*

$$f(x, y) = f_X(x) f_Y(y)$$

*for all $x, y \in \mathbb{R}$. Furthermore and in general, if $f : \mathbb{R}^n \to \mathbb{R}$ is the joint density of finitely many (discrete or continuous) random variables $X_1, \ldots, X_n : S \to \mathbb{R}$, then $X_1, \ldots, X_n$ are independent if and only if*

$$f(x_1, \ldots, x_n) = \prod_{k=1}^{n} f_{X_k}(x_k)$$

*for all $(x_1, \ldots, x_n) \in \mathbb{R}^n$.*

*Proof.* We will prove only the former statement (for the case $n = 2$) and leave the completely analogous but somewhat more cumbersome proof of the latter as an exercise to the reader. So let us assume that $f$ is the density of two independent random variables $X, Y : S \to \mathbb{R}$. If $X$ and $Y$ are discrete, then

$$f(x, y) = P(X = x \wedge Y = y) = P(\{s \in S \mid X(s) = x\} \cap \{s \in S \mid Y(s) = y\})$$
$$= P(\{s \in S \mid X(s) \in \{x\}\} \cap \{s \in S \mid Y(s) \in \{y\}\})$$
$$= P(\{s \in S \mid X(s) \in \{x\}\}) P(\{s \in S \mid Y(s) \in \{y\}\})$$
$$= P(\{X = x \wedge Y \in \mathbb{R}\}) P(\{X \in \mathbb{R} \wedge Y = y\})$$
$$= \left(\sum_{y \in R(Y)} f(x, y)\right) \left(\sum_{x \in R(X)} f(x, y)\right) = f_X(x) f_Y(y),$$

and if $X$ and $Y$ are continuous (and if $f$ is continuous as well—in the standard calculus sense), then the cumulative density is

$$F(x,y) = \int_{-\infty}^{y}\int_{-\infty}^{x} f(u,v)\,du\,dv = P(X \leq x \wedge Y \leq Y)$$
$$= P(\{s \in S \mid X(s) \in (-\infty,x]\} \cap \{s \in S \mid Y(s) \in (-\infty,y]\})$$
$$= P(\{s \in S \mid X(s) \in (-\infty,x]\})P(\{s \in S \mid Y(s) \in (-\infty,y]\})$$
$$= P(\{X \in (-\infty,x] \wedge Y \in \mathbb{R}\})P(\{X \in \mathbb{R} \wedge Y \in (-\infty,y]\})$$
$$= \left(\int_{-\infty}^{x}\int_{-\infty}^{\infty} f(u,v)\,dv\,du\right)\left(\int_{-\infty}^{y}\int_{-\infty}^{\infty} f(u,v)\,du\,dv\right).$$

Taking the derivative with respect to both $x$ and $y$ yields

$$f(x,y) = \left(\int_{-\infty}^{\infty} f(x,v)\,dv\right)\left(\int_{-\infty}^{\infty} f(u,y)\,du\right) = f_X(x)f_Y(y),$$

as desired. Alternatively, we can also use (2.38) in conjunction with (2.15) to argue—somewhat informally—that

$$f(x,y)\,dx\,dy = P(X \in [x,x+dx] \wedge Y \in [y,y+dy])$$
$$= P(\{s \in S \mid X(s) \in [x,x+dx]\} \cap \{s \in S \mid Y(s) \in [y,y+dy]\})$$
$$= P(\{s \in S \mid X(s) \in [x,x+dx]\})P(\{s \in S \mid Y(s) \in [y,y+dy]\})$$
$$= P(X \in [x,x+dx])P(Y \in [y,y+dy]) = f_X(x)\,dx\,f_Y(y)\,dy,$$

and therefore,

$$f(x,y) = f_X(x)f_Y(y).$$

Conversely, if

$$f(x,y) = f_X(x)f_Y(y)$$

for all $x,y \in \mathbb{R}$ and if $X$ and $Y$ are discrete and $A,B \subset \mathbb{R}$, then

$$P(\{X \in A\} \cap \{Y \in B\}) = P(\{X \in (A \cap R(X)) \wedge Y \in (B \cap R(Y)\})$$
$$= \sum_{x \in A \cap R(X)} \sum_{y \in B \cap R(Y)} f(x,y)$$
$$= \sum_{x \in A \cap R(X)} \sum_{y \in B \cap R(Y)} f_X(x)f_Y(y)$$
$$= \left(\sum_{x \in A \cap R(X)} f_X(x)\right)\left(\sum_{y \in B \cap R(Y)} f_Y(y)\right)$$
$$= P(\{X \in A\})P(\{Y \in B\}),$$

and similarly, if $X$ and $Y$ are continuous and if $A$ and $B$ are suitable subsets of $\mathbb{R}$, then

$$P(\{X \in A\} \cap \{Y \in B\}) = P(\{X \in A \wedge Y \in B\}) = P(\{(X, Y) \in A \times B\})$$

$$= \int_{A \times B} f(x, y) \, dx \, dy = \int_B \int_A f(x, y) \, dx \, dy$$

$$= \int_B \int_A f_X(x) f_Y(y) \, dx \, dy = \left( \int_A f_X(x) \, dx \right) \left( \int_B f_Y(y) \, dy \right)$$

$$= P(\{X \in A\}) P(\{Y \in B\}).$$

Consequently, $X$ and $Y$ are independent. □

*Remark.* The theoretical limitations that this introductory text is naturally subject to did not allow us to state the proof of the preceding theorem in a fully adequate form. For it was due these limitations that we imposed the additional assumption that $f$ be continuous in the first part of the proof and that we used the identity

$$\int_{A \times B} f(x, y) \, dx \, dy = \int_B \int_A f(x, y) \, dx \, dy$$

in the latter part of the proof without a proper explanation. As a matter of course, this latter identity is routinely employed in multivariable calculus in the special case where $A$ and $B$ are intervals, but to establish it in this present exposition in a fully rigorous manner and in full generality is unfortunately not possible.

**2.9.4 Example.** Given the marginal densities

$$f_X(x) = \begin{cases} 1 & \text{if } x \in (0, 1], \\ 0 & \text{otherwise} \end{cases}$$

and

$$f_Y(y) = \begin{cases} -\ln(y) & \text{if } y \in (0, 1], \\ 0 & \text{otherwise} \end{cases}$$

that we found in Example 2.8.10, it follows that the product

$$f_X(x) f_Y(y) = \begin{cases} -\ln(y) & \text{if } x, y \in (0, 1], \\ 0 & \text{otherwise} \end{cases}$$

is not equal to the corresponding joint density

$$f(x, y) = \begin{cases} 1/x & \text{if } 0 < y \leq x \leq 1, \\ 0 & \text{otherwise.} \end{cases}$$

Thus $X$ and $Y$ are not independent.

**2.9.5 Example.** If the joint density of two random variables $X, Y : S \to \mathbb{R}$ is

$$f(x, y) = \begin{cases} 6x^2 y & \text{if } (x, y) \in [0, 1] \times [0, 1], \\ 0 & \text{otherwise,} \end{cases}$$

then

$$f_X(x) = \int_{-\infty}^{\infty} f(x, y)\, dy = \int_0^1 6x^2 y\, dy = 3x^2$$

and

$$f_Y(y) = \int_0^1 6x^2 y\, dx = 2y.$$

Hence

$$f(x, y) = f_X(x) f_Y(y)$$

and, by implication, $X$ and $Y$ are independent.

**2.9.6 Example.** Let $X(s, t) := t + s$ and $Y(s, t) := t - s$ for all $(s, t) \in \mathbb{R}^2$ and assume that the probability assigned to an event $E \subset \mathbb{R}^2$ is $P(E) = \int_E p(s, t)\, ds\, dt$ where

$$p(s, t) := \frac{e^{-(s^2 + t^2)/2}}{2\pi}$$

for all $(s, t) \in \mathbb{R}^2$. We wish to show that $X$ and $Y$ are independent. To do so, we first observe that the cumulative density of $X$ and $Y$ is

$$F(x, y) = P(X \leq x \wedge Y \leq y) = P(\{(s, t) \in \mathbb{R}^2 \mid s + t \leq x \wedge t - s \leq y\})$$
$$= P(\{(s, t) \in \mathbb{R}^2 \mid t \leq x - s \wedge t \leq y + s\}).$$

Since the equations $t = x - s$ and $t = y + s$ describe two lines in the $st$-plane (with slopes $-1$ and $1$, respectively) that intersect in the point

$$\left( \frac{x - y}{2}, \frac{x + y}{2} \right),$$

it follows that

$$F(x, y) = \int_{-\infty}^{(x-y)/2} \int_{-\infty}^{s+y} p(s, t)\, dt\, ds + \int_{(x-y)/2}^{\infty} \int_{-\infty}^{x-s} p(s, t)\, dt\, ds. \quad (2.44)$$

In order to write this equation in a more convenient form, we introduce the coordinate transformation

$$\mathbf{g}(\sigma, \tau) := \begin{pmatrix} \cos(\pi/4) & -\sin(\pi/4) \\ \sin(\pi/4) & \cos(\pi/4) \end{pmatrix} \begin{pmatrix} \sigma \\ \tau \end{pmatrix} + \begin{pmatrix} (x - y)/2 \\ (x + y)/2 \end{pmatrix}$$

$$= \frac{1}{\sqrt{2}} \begin{pmatrix} 1 & -1 \\ 1 & 1 \end{pmatrix} \begin{pmatrix} \sigma + x/\sqrt{2} \\ \tau + y/\sqrt{2} \end{pmatrix} =: \mathbf{A} \begin{pmatrix} \sigma + x/\sqrt{2} \\ \tau + y/\sqrt{2} \end{pmatrix}$$

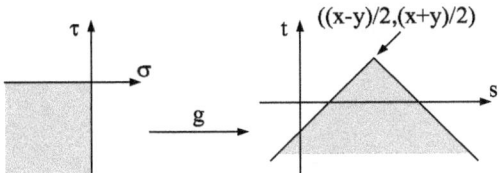

Figure 2.14: the transformation **g**.

which maps the third quadrant in the $xy$-plane—by means of a rotation and a translation—onto the region of integration covered by the integrals in (2.44) (Figure 2.14). Since neither a rotation nor a translation distorts area content, it follows that the absolute value of the Jacobian determinant of **g** must be equal to one. The following direct calculation confirms this claim:

$$|J_\mathbf{g}| = \left| \det \begin{pmatrix} \partial g_1/\partial \sigma & \partial g_1/\partial \tau \\ \partial g_2/\partial \sigma & \partial g_2/\partial \tau \end{pmatrix} \right| = |\det(\mathbf{A})| = 1.$$

Hence

$$F(x,y) = \int_{-\infty}^0 \int_{-\infty}^0 p(\mathbf{g}(\sigma,\tau))|J_\mathbf{g}(\sigma,\tau)|\, d\sigma\, d\tau = \int_{-\infty}^0 \int_{-\infty}^0 p(\mathbf{g}(\sigma,\tau))\, d\sigma\, d\tau.$$

Since the matrix **A** describes a rotation in the plane (by 45° counterclockwise), we may infer that

$$s^2 + t^2 = \left\| \begin{pmatrix} s \\ t \end{pmatrix} \right\|^2 = \|\mathbf{g}(\sigma,\tau)\|^2 = \left\| \mathbf{A} \begin{pmatrix} \sigma + x/\sqrt{2} \\ \tau + y/\sqrt{2} \end{pmatrix} \right\|^2 = \left\| \begin{pmatrix} \sigma + x/\sqrt{2} \\ \tau + y/\sqrt{2} \end{pmatrix} \right\|^2$$

$$= \left( \sigma + \frac{x}{\sqrt{2}} \right)^2 + \left( \tau + \frac{y}{\sqrt{2}} \right)^2,$$

and therefore,

$$p(\mathbf{g}(\sigma,\tau)) = \frac{e^{-\left((\sigma+x/\sqrt{2})^2 + (\tau+y/\sqrt{2})^2\right)/2}}{2\pi}.$$

Thus

$$f(x,y) = \frac{\partial^2}{\partial x \partial y} F(x,y) = \frac{\partial^2}{\partial x \partial y} \int_{-\infty}^0 \int_{-\infty}^0 \frac{e^{-\left((\sigma+x/\sqrt{2})^2 + (\tau+y/\sqrt{2})^2\right)/2}}{2\pi}\, d\sigma\, d\tau$$

$$= \frac{\partial}{\partial x} \int_{-\infty}^0 \frac{e^{-(\sigma+x/\sqrt{2})^2/2}}{\sqrt{2\pi}}\, d\sigma \, \frac{\partial}{\partial y} \int_{-\infty}^0 \frac{e^{-(\tau+y/\sqrt{2})^2/2}}{\sqrt{2\pi}}\, d\tau$$

$$= \int_{-\infty}^0 \frac{-(\sigma+x/\sqrt{2})e^{-(\sigma+x/\sqrt{2})^2/2}}{\sqrt{2\pi}\sqrt{2}}\, d\sigma \int_{-\infty}^0 \frac{-(\tau+y/\sqrt{2})e^{-(\tau+y/\sqrt{2})^2/2}}{\sqrt{2\pi}\sqrt{2}}\, d\tau$$

$$= \frac{e^{-(\sigma + x/\sqrt{2})^2/2}}{\sqrt{2\pi}\sqrt{2}} \Big|_{-\infty}^{0} \; \frac{e^{-(\tau + y/\sqrt{2})^2/2}}{\sqrt{2\pi}\sqrt{2}} \Big|_{-\infty}^{0}$$

$$= \frac{e^{-x^2/4}}{\sqrt{2\pi}\sqrt{2}} \cdot \frac{e^{-y^2/4}}{\sqrt{2\pi}\sqrt{2}},$$

and, by implication,

$$f_X(x) = \frac{e^{-x^2/4}}{\sqrt{2\pi}\sqrt{2}} \int_{-\infty}^{\infty} \frac{e^{-y^2/4}}{\sqrt{2\pi}\sqrt{2}} \, dy = \frac{e^{-x^2/4}}{\sqrt{2\pi}\sqrt{2}}.$$

Similarly, we also find that

$$f_Y(y) = \frac{e^{-y^2/4}}{\sqrt{2\pi}\sqrt{2}},$$

and this shows that $X$ and $Y$ are indeed independent because $f(x,y) = f_X(x)f_Y(y)$.

*Remark.* An alternative derivation of the result in the preceding example can be given as follows: starting from the equation

$$F(x,y) = \int_{-\infty}^{(x-y)/2} \int_{-\infty}^{s+y} p(s,t) \, dt \, ds + \int_{(x-y)/2}^{\infty} \int_{-\infty}^{x-s} p(s,t) \, dt \, ds$$

and using the fundamental theorem of calculus in conjunction with the fact that

$$\frac{\partial^2}{\partial x \partial y} F(x,y) = \frac{\partial^2}{\partial y \partial x} F(x,y),$$

it is easy to see that

$$f(x,y) =$$
$$= \frac{1}{2} \frac{\partial}{\partial y} \int_{-\infty}^{(x-y)/2+y} p\left(\frac{x-y}{2},t\right) dt + \frac{1}{2} \frac{\partial}{\partial x} \int_{-\infty}^{x-(x-y)/2} p\left(\frac{x-y}{2},t\right) dt$$
$$= \frac{1}{2} \frac{\partial}{\partial y} \int_{-\infty}^{(x+y)/2} p\left(\frac{x-y}{2},t\right) dt + \frac{1}{2} \frac{\partial}{\partial x} \int_{-\infty}^{(x+y)/2} p\left(\frac{x-y}{2},t\right) dt.$$

Now let $Q$ be an antiderivative of $p$ with respect to $t$, that is, $\partial Q/\partial t = p$, and set $R(s) := \lim_{t \to -\infty} Q(s,t)$. Then

$$f(x,y) = \frac{1}{2} \frac{\partial}{\partial y} \left( Q\left(\frac{x-y}{2}, \frac{x+y}{2}\right) - R\left(\frac{x-y}{2}\right) \right)$$
$$+ \frac{1}{2} \frac{\partial}{\partial x} \left( Q\left(\frac{x-y}{2}, \frac{x+y}{2}\right) - R\left(\frac{x-y}{2}\right) \right)$$

$$= \frac{1}{2} \nabla Q \left( \frac{x-y}{2}, \frac{x+y}{2} \right) \cdot \begin{pmatrix} -1/2 \\ 1/2 \end{pmatrix} + \frac{1}{4} R' \left( \frac{x-y}{2} \right)$$

$$+ \frac{1}{2} \nabla Q \left( \frac{x-y}{2}, \frac{x+y}{2} \right) \cdot \begin{pmatrix} 1/2 \\ 1/2 \end{pmatrix} - \frac{1}{4} R' \left( \frac{x-y}{2} \right)$$

$$= \frac{1}{2} \left. \begin{pmatrix} \partial Q/\partial s \\ p \end{pmatrix} \right|_{((x-y)/2,(x+y)/2)} \cdot \begin{pmatrix} -1/2 \\ 1/2 \end{pmatrix}$$

$$+ \frac{1}{2} \left. \begin{pmatrix} \partial Q/\partial s \\ p \end{pmatrix} \right|_{((x-y)/2,(x+y)/2)} \cdot \begin{pmatrix} 1/2 \\ 1/2 \end{pmatrix}$$

$$= \frac{1}{2} p \left( \frac{x-y}{2}, \frac{x+y}{2} \right) = \frac{e^{-((x-y)^2+(x+y)^2)/8}}{4\pi} = \frac{e^{-(x^2+y^2)/4}}{4\pi}$$

$$= \frac{e^{-x^2/4}}{\sqrt{2\pi}\sqrt{2}} \cdot \frac{e^{-y^2/4}}{\sqrt{2\pi}\sqrt{2}},$$

and as before, we find that

$$f_X(x) = \frac{e^{-x^2/4}}{\sqrt{2\pi}\sqrt{2}} \int_{-\infty}^{\infty} \frac{e^{-y^2/4}}{\sqrt{2\pi}\sqrt{2}} \, dy = \frac{e^{-x^2/4}}{\sqrt{2\pi}\sqrt{2}},$$

and similarly, that

$$f_Y(y) = \frac{e^{-y^2/4}}{\sqrt{2\pi}\sqrt{2}}.$$

**2.9.7 Example.** Let us assume that in a certain population of American adults that are eligible to vote gender and political party affiliation are distributed as follows:

|             | women | men |
|-------------|-------|-----|
| Democrats   | 28%   | 21% |
| Republicans | 24%   | 27% |

This table suggests that gender and political party affiliation are not independent because women are more likely to vote democratic than men. In order to formally verify this observation, we use the initial letters $W$, $M$, $D$, and $R$ to denote the words 'woman', 'man', 'Democrat', and 'Republican', respectively, and set

$$S := \{(W, D), (W, R), (M, D), (M, R)\}.$$

Given this sample space, it is natural to represent gender and party affiliation by the following random variables:

$$X(W, D) := 1, \ X(W, R) := 1, \ X(M, D) := 0, \ X(M, R) := 0$$

and

$$Y(W, D) := 1, \ Y(W, R) := 0, \ Y(M, D) := 1, \ Y(M, R) := 0.$$

According to the table above, these definitions imply that the discrete joint density of $X$ and $Y$ is

$$f(x,y) = \begin{cases} 0.28 & \text{if } (x,y) = (1,1), \\ 0.24 & \text{if } (x,y) = (1,0), \\ 0.21 & \text{if } (x,y) = (0,1), \\ 0.27 & \text{if } (x,y) = (0,0), \\ 0 & \text{otherwise.} \end{cases} \qquad (2.45)$$

Consequently, the corresponding marginal densities are

$$f_X(x) = \begin{cases} f(1,1) + f(1,0) = 0.52 & \text{if } x = 1, \\ f(0,1) + f(0,0) = 0.48 & \text{if } x = 0, \\ 0 & \text{otherwise} \end{cases}$$

and

$$f_Y(y) = \begin{cases} f(1,1) + f(0,1) = 0.49 & \text{if } y = 1, \\ f(1,0) + f(0,0) = 0.51 & \text{if } y = 0, \\ 0 & \text{otherwise.} \end{cases}$$

This shows that $X$ and $Y$ are not independent, because the product

$$f_X(x) f_Y(y) = \begin{cases} 0.52 \cdot 0.49 = 0.2548 & \text{if } (x,y) = (1,1) \\ 0.52 \cdot 0.51 = 0.2652 & \text{if } (x,y) = (1,0) \\ 0.48 \cdot 0.49 = 0.2352 & \text{if } (x,y) = (0,1) \\ 0.48 \cdot 0.51 = 0.2448 & \text{if } (x,y) = (0,0) \\ 0 & \text{otherwise} \end{cases}$$

is evidently not equal to the joint density in (2.45). Note: a more detailed analysis of this example that shows the female gender to be positively correlated with the Democratic party affiliation will be provided below, in Example 2.11.14.

To proceed we state and prove two very important properties concerning expected values and variances of products and sums of independent random variables:

**2.9.8 Theorem.** *If $X_1, \ldots, X_n : S \to \mathbb{R}$ are independent random variables, then*

$$E(X_1 \cdots X_n) = E(X_1) \cdots E(X_n)$$

*and*

$$\mathrm{Var}(X_1 + \cdots + X_n) = \mathrm{Var}(X_1) + \cdots + \mathrm{Var}(X_n).$$

*Proof.* Since the general case of $n$ random variables is completely analogous to the case of only two random variables $X$ and $Y$, we will be content to consider only the latter. If $X$ and $Y$ are independent and discrete with joint density $f(x, y)$, then

$$E(XY) = \sum_{x \in R(X)} \sum_{y \in R(Y)} xyf(x, y) = \sum_{x \in R(X)} \sum_{y \in R(Y)} xyf_X(x)f_Y(y)$$

$$= \left( \sum_{x \in R(X)} xf_X(x) \right) \left( \sum_{y \in R(Y)} yf_Y(y) \right) = E(X)E(Y),$$

and similarly, if $X$ and $Y$ are continuous, then

$$E(XY) = \int_{-\infty}^{\infty} \int_{-\infty}^{\infty} xyf(x, y)\, dx\, dy = \int_{-\infty}^{\infty} \int_{-\infty}^{\infty} xyf_X(x)f_Y(y)\, dx\, dy$$

$$= \left( \int_{-\infty}^{\infty} xf_X(x)\, dx \right) \left( \int_{-\infty}^{\infty} yf_Y(y)\, dy \right) = E(X)E(Y).$$

Furthermore, this result allows us to infer that

$$\begin{aligned}
\mathrm{Var}(X + Y) &= E((X + Y)^2) - E(X + Y)^2 \\
&= E(X^2 + 2XY + Y^2) - (E(X) + E(Y))^2 \\
&= E(X^2) + 2E(XY) + E(Y^2) \\
&\quad - E(X)^2 - 2E(X)E(Y) - E(Y)^2 \\
&= E(X^2) - E(X)^2 + E(Y^2) - E(Y)^2 \\
&= \mathrm{Var}(X) + \mathrm{Var}(Y),
\end{aligned}$$

as desired. □

*Remark.* It is important to understand that Theorem 2.9.8 asserts a one-way implication rather than a two-way equivalence: *if $X$ and $Y$ are independent, then $E(XY) = E(X)E(Y)$* (in the case where $n = 2$), but the reverse conclusion that $X$ and $Y$ are independent whenever $E(XY) = E(X)E(Y)$ is in general not valid. For instance, if $S = \{a, b, c, d\}$ is a four-element sample space with equal outcome probabilities $P(a) = P(b) = P(c) = P(d) = 1/4$ and if

$$X(a) := 1, \ X(b) := -1, \ X(c) := 0, \ X(d) := 0,$$

and

$$Y(a) := 0, \ Y(b) := 0, \ Y(c) := 1, \ Y(d) := -1,$$

then

$$X(s)Y(s) = 0$$

for all $s \in S$ and

$$E(X) = 1 \cdot \frac{1}{4} - 1 \cdot \frac{1}{4} = 0 = E(Y).$$

Hence

$$E(XY) = E(0) = 0 = E(X)E(Y),$$

but $X$ and $Y$ are not independent, because

$$P(\{X = 1\} \cap \{Y = 1\}) = P(\{a\} \cap \{c\}) = P(\emptyset) = 0$$

$$\neq \frac{1}{16} = P(\{a\})P(\{b\}) = P(\{X = 1\})P(\{Y = 1\}).$$

Furthermore, since the proof of Theorem 2.9.8 shows that $Var(X + Y) = Var(X) + Var(Y)$ whenever $E(XY) = E(X)E(Y)$, it follows that any non-independent random variables $X$ and $Y$ that satisfy the latter equation will also satisfy the former. Consequently, the example above shows that there do exists non-independent random variables $X$ and $Y$ for which both equations in Theorem 2.9.8 are satisfied (in the case where $n = 2$).

**2.9.9 Example.** Let $S_n$ be the set of all sequences of length $n$ in the letters $H$ and $T$ and let $X_n : S \to \mathbb{R}$ be the corresponding binomial random variable with parameters $n$ and $p$ (i.e., $p = P(H)$ and $X_n(s) := \#$ of letters $H$ in $s$). For a given $s \in S_n$ and a given $k \in \{1, \ldots, n\}$ we set (as in Example 2.3.15)

$$X_{n,k}(s) := \begin{cases} 1 & \text{if the } k\text{-th letter in } s \text{ is an } H, \\ 0 & \text{if the } k\text{-th letter in } s \text{ is a } T, \end{cases}$$

and

$$\{0, 1\}^n = \{(x_1, \ldots, x_n) \in \mathbb{R}^n \mid x_i \in \{0, 1\} \text{ for all } i \in \{1, \ldots, n\}\}.$$

Then the joint discrete density of the random variables $X_{n,1}, \ldots, X_{n,n}$ is

$$f_n(x_1, \ldots, x_n) = \begin{cases} p^{\sum_{k=1}^{n} x_k}(1-p)^{n - \sum_{k=1}^{n} x_k} & \text{if } (x_1, \ldots, x_n) \in \{0, 1\}^n, \\ 0 & \text{otherwise}, \end{cases}$$

and for every $k \in \{1, \ldots, n\}$, the marginal density of $X_{n,k}$ is

$$f_{n,X_{n,k}}(x_k) = \sum_{(u_1, \ldots, u_{k-1}, u_{k+1}, \ldots, u_n) \in \{0,1\}^{n-1}} f_n(u_1, \ldots, u_{k-1}, x_k, u_{k+1}, \ldots, u_n)$$

$$= p^{x_k}(1-p)^{1-x_k} \sum_{i=0}^{n-1} \binom{n-1}{i} p^i (1-p)^{n-1-i} = p^{x_k}(1-p)^{1-x_k}$$

whenever $x_k \in \{0, 1\}$ and

$$f_{n,X_{n,k}}(x_k) = 0$$

whenever $x_k \in \mathbb{R} \setminus \{0,1\}$. Consequently,

$$\prod_{k=1}^{n} f_{n,X_{n,k}}(x_k) = \begin{cases} \prod_{k=1}^{n} p^{x_k}(1-p)^{1-x_k} & \text{if } (x_1,\ldots,x_n) \in \{0,1\}^n, \\ 0 & \text{otherwise} \end{cases}$$

$$= \begin{cases} p^{\sum_{k=1}^{n} x_k}(1-p)^{n-\sum_{k=1}^{n} x_k} & \text{if } (x_1,\ldots,x_n) \in \{0,1\}^n, \\ 0 & \text{otherwise} \end{cases}$$

$$= f_n(x_1,\ldots,x_n),$$

and therefore, $X_{n,1},\ldots,X_{n,n}$ are independent for all $n \in \mathbb{N}$. Furthermore, since

$$X_n(s) = \sum_{k=1}^{n} X_{n,k}(s)$$

for all $s \in S_n$, it follows that

$$E(X_n) = \sum_{k=1}^{n} E(X_{n,k}) = \sum_{k=1}^{n}(1 \cdot p + 0 \cdot (1-p)) = np \quad \text{(by Theorem 2.3.5)}$$

and

$$\text{Var}(X_n) = \sum_{k=1}^{n} \text{Var}(X_{n,k}) \quad \text{(by Theorem 2.9.8)}.$$

$$= \sum_{k=1}^{n}(E(X_{n,k}^2) - E(X_{n,k})^2) = \sum_{k=1}^{n}(1^2 \cdot p + 0^2 \cdot (1-p) - p^2)$$

$$= np(1-p).$$

Thus we have confirmed the results that we found earlier in Examples 2.3.3 and 2.3.10.

**2.9.10 Example.** Using the results of the preceding example in conjunction with Theorem 2.3.7b, we also can confirm the limit property that we previously established in Example 2.3.15:

$$\text{Var}\left(\frac{1}{n}\sum_{k=1}^{n} X_{n,k}\right) = \frac{1}{n^2}\text{Var}\left(\sum_{k=1}^{n} X_{n,k}\right) = \frac{np(1-p)}{n^2} = \frac{p(1-p)}{n},$$

and therefore,

$$\lim_{n\to\infty} \text{Var}\left(\frac{1}{n}\sum_{k=1}^{n} X_{n,k}\right) = 0.$$

A generalization of this limit property can be stated as follows:

**2.9.11 Theorem.** *If $(S_n)_{n\in\mathbb{N}}$ is a sequence of sample spaces and if for every $n \in \mathbb{N}$ we are given independent random variables*

$$X_{n,1}, \ldots, X_{n,n} : S_n \to \mathbb{R}$$

*with common variance $\sigma^2$ (that is, $\mathrm{Var}(X_{n,k}) = \sigma^2$ for all $n \in \mathbb{N}$ and all $k \in \{1, \ldots, n\}$), then*

$$\lim_{n\to\infty} \mathrm{Var}\left(\frac{1}{n}\sum_{k=1}^{n} X_{n.k}\right) = 0.$$

*Proof.* Using Theorem 2.3.7b in conjunction with Theorem 2.9.8, we may infer that

$$\lim_{n\to\infty} \mathrm{Var}\left(\frac{1}{n}\sum_{k=1}^{n} X_{n,k}\right) = \lim_{n\to\infty} \frac{1}{n^2}\mathrm{Var}\left(\sum_{k=1}^{n} X_{n,k}\right) = \lim_{n\to\infty} \frac{1}{n^2}\sum_{k=1}^{n}\mathrm{Var}(X_{n,k})$$

$$= \lim_{n\to\infty} \frac{1}{n^2}\sum_{k=1}^{n}\sigma^2 = \lim_{n\to\infty}\frac{\sigma^2}{n} = 0,$$

as desired.                                                                    $\square$

To conclude this section, we will now prove that the sum of independent normal random variables is always normal as well.

**2.9.12 Theorem.** *If $X_1, \ldots, X_n : S \to \mathbb{R}$ are independent normal random variables such that $E(X_k) = \mu_k$ and $\mathrm{Var}(X_k) = \sigma_k^2$ for all $k \in \{1, \ldots, n\}$, then $X := X_1 + \cdots + X_n$ is normal as well with mean $\mu = \mu_1 + \cdots + \mu_n$ and variance $\sigma^2 = \sigma_1^2 + \cdots + \sigma_n^2$.*

*Proof.* Since the general case of $n$ independent normal random variables is an immediate consequence of the special case $n = 2$ (simply apply the statement for $n = 2$ repeatedly to $X_1 + X_2$, $(X_1 + X_2) + X_3$ etc.) and since the statements concerning the mean and variance of $X$ are direct consequences of the preceding Theorem 2.9.8, it is sufficient to demonstrate that the sum of two independent normal random variables $X, Y : S \to \mathbb{R}$ is normal as well. Denoting by $\mu$, $\nu$, $\sigma^2$, and $\tau^2$ the respective means and variances of $X$ and $Y$, it follows that the corresponding marginal density functions are

$$f_X(x) = \frac{e^{-(x-\mu)^2/2\sigma^2}}{\sqrt{2\pi}\sigma} \text{ and } f_Y(y) = \frac{e^{-(y-\nu)^2/2\tau^2}}{\sqrt{2\pi}\tau}.$$

Since $X$ and $Y$ are independent, it further follows that the joint density of $X$ and $Y$ is

$$f(x,y) = f_X(x)f_Y(y) = \frac{e^{-(x-\mu)^2/2\sigma^2 - (y-\nu)^2/2\tau^2}}{2\pi\sigma\tau}.$$

Consequently, the cumulative density of $X + Y$ is

$$F(z) = P(X + Y \le z) = P(Y \le z - X) = \int_{-\infty}^{\infty} \int_{-\infty}^{z-x} f_X(x) f_Y(y)\, dy\, dx$$

and, by implication,

$$
\begin{aligned}
f(z) = F'(z) &= \frac{d}{dz} \int_{-\infty}^{\infty} \int_{-\infty}^{z-x} f_X(x) f_Y(y)\, dy\, dx \\
&= \int_{-\infty}^{\infty} f_X(x) \left( \frac{d}{dz} \int_{-\infty}^{z-x} f_Y(y)\, dy \right) dx \\
&= \int_{-\infty}^{\infty} f_X(x) f_Y(z - x)\, dx.
\end{aligned}
$$

Using elementary algebra, it is easy to see that

$$
\begin{aligned}
f_X(x) f_Y(z - x) &= \frac{e^{-(x-\mu)^2/2\sigma^2 - (z-x-\nu)^2/2\tau^2}}{2\pi\sigma\tau} \\
&= \frac{e^{-(\sigma^2+\tau^2)(x - (\tau^2\mu + \sigma^2(z-\nu))/(\sigma^2+\tau^2))^2/(2\sigma^2\tau^2)} e^{-(z-(\nu+\mu))^2/(2(\sigma^2+\tau^2))}}{2\pi\sigma\tau},
\end{aligned}
$$

and therefore,

$$
\begin{aligned}
f(z) &= \\
&= \frac{e^{-(z-(\nu+\mu))^2/(2(\sigma^2+\tau^2))}}{2\pi\sigma\tau} \int_{-\infty}^{\infty} e^{-(\sigma^2+\tau^2)(x-(\tau^2\mu+\sigma^2(z-\nu))/(\sigma^2+\tau^2))^2/(2\sigma^2\tau^2)}\, dx \\
&= \frac{e^{-(z-(\nu+\mu))^2/(2(\sigma^2+\tau^2))} \sqrt{2\pi}\sigma\tau}{2\pi\sigma\tau\sqrt{\sigma^2+\tau^2}} = \frac{e^{-(z-(\nu+\mu))^2/(2(\sigma^2+\tau^2))}}{\sqrt{2\pi}\sqrt{\sigma^2+\tau^2}}.
\end{aligned}
$$

Thus, we arrive at the desired conclusion that the density of the random variable $X + Y$ is normal with mean $\mu + \nu$ and variance $\sigma^2 + \tau^2$. $\qquad\square$

## Exercises

**2.9.13.** Decide whether the random variables $X$ and $Y$ in Exercise 2.8.25 are independent.

**2.9.14.** Let $S := \{1, \ldots, n\} \times \{1, \ldots, n\}$ and let $P(\{(i,j)\}) := 1/n^2$ for all $(i, j) \in S$. Are the random variables $X(i,j) := i$ and $Y(i,j) := j$ (defined on $S$) independent?

**2.9.15.** Assume that a point $(s, t)$ is selected at random from the set $[-1, 0] \times [-1, 0] \cup [0, 1] \times [0, 1]$ (all points are equally likely to be chosen). Find the joint density of $X(s, t) := s$ and $Y(s, t) := t$ and decide whether $X$ and $Y$ are independent.

**2.9.16.** Assume that $X, Y, Z : S \to \mathbb{R}$ are independent random variables that are uniformly distributed over the interval $[0, 1]$, that is, the density of each of them is

$$f(x) = \begin{cases} 1 & \text{if } x \in [0, 1], \\ 0 & \text{otherwise.} \end{cases}$$

a) Find the joint density of $X$, $Y$, and $Z$.

b) Find the cumulative density and the regular density of the product $XYZ$.

c) Use your answer to b) to find the expected value of $XYZ$, and then verify your result by using the density determined in a).

**2.9.17.** Assume that $X$ and $Y$ are independent random variables with respective densities $f(x)$ and $g(y)$. Find the joint density of $X^2$ and $Y^2$ and show that $X^2$ and $Y^2$ are independent as well.

**2.9.18.** Assume that $X$ and $Y$ are independent random variables with respective means $\mu_X$ and $\mu_Y$ and respective standard deviations $\sigma_X$ and $\sigma_Y$. Use the result of Exercise 2.9.17 to find the variance of $XY$ (in terms of $\mu_X$, $\mu_Y$, $\sigma_X$ and $\sigma_Y$).

**2.9.19.** Let $\lambda \in \mathbb{R} \setminus \{0\}$ and assume that $X$ is a normal random variable with mean $\mu$ and variance $\sigma^2$. Show that $\lambda X$ is normal with mean $\lambda \mu$ and variance $\lambda^2 \sigma^2$.

**2.9.20.** Assume that $X$ and $Y$ are independent normal random variables with $\mu_X = 2$, $\sigma_X = 4$, $\mu_Y = 1$, and $\sigma_Y = 2$. Use the result of Exercise 2.9.19 to determine $P(|2X - 3Y| < 10)$.

**2.9.21.** Assume that $X_1, \ldots, X_n$ are independent random variables with common mean $\mu$ and common standard deviation $\sigma$. Show that $\lambda_1 X_1, \ldots, \lambda_n X_n$ are independent as well for all $\lambda_1, \ldots, \lambda_n \in \mathbb{R} \setminus \{0\}$.

**2.9.22.** Assume that $X_1, \ldots, X_n$ are independent random variables with common mean $\mu$ and common standard deviation $\sigma$. Use Exercise 2.9.21 to find the mean and the standard deviation of $\lambda_1 X_1 + \cdots + \lambda_n X_n$.

**2.9.23.** Assume that $X$ and $Y$ are independent normal random variables such that $E(X) = E(Y) = 0$, $\sigma(X) = \sigma(Y) = 4$. Use this information to find the probability $P(X^2 + Y^2 \leq 1)$.

**2.9.24.** Assume that $X$ and $Y$ are independent normal random variables such that $E(X) = E(Y) = 1$ and $\sigma(X) = \sigma(Y) = \sqrt{9/2}$. Use the 68-95-99.7 rule to find an estimate for the probability $P(X - Y \leq 6)$

**2.9.25.** Assume that the joint density of two random variables $X$ and $Y$ is

$$f(x, y) = \begin{cases} C/(x + y) & \text{if } (x, y) \in [1, 2] \times [1, 2], \\ 0 & \text{otherwise.} \end{cases}$$

**a)** Find the value of $C$.

**b)** Find $E(X)$, $E(Y)$, and $E(XY)$.

**c)** Are $X$ and $Y$ independent?

**2.9.26.** Assume that the joint density of two random variables $X$ and $Y$ is

$$f(x, y) = \begin{cases} C/(xy) & \text{if } (x, y) \in [1, e] \times [1, e], \\ 0 & \text{otherwise.} \end{cases}$$

**a)** Find the value of $C$.

**b)** Find $E(X)$, $E(Y)$, and $E(XY)$.

**c)** Are $X$ and $Y$ independent?

**2.9.27.** Let the sample space $S := [0, 1]$ be equipped with the uniform density and let two random variables $X, Y : S \rightarrow \mathbb{R}$ be defined by the equations $X(s) := s$ and $Y(s) := s - s^2$.

**a)** Find $E(X)$, $E(Y)$, and $E(XY)$.

**b)** Are $X$ and $Y$ independent?

**2.9.28.** Assume that $X$ and $Y$ are independent standard normal random variables. Find the values of $P(Y > X + 1)$ and $P(|Y - X| < 1)$.

**2.9.29.** Assume that $X$ and $Y$ are independent standard normal random variables. Find the density of $Z := X^2 + Y^2$.

## 2.10 Ideal Gases and Statistical Mechanics

In order to demonstrate how the concepts developed in the preceding sections can be used to describe physical reality, we will study in this present section a statistical model of particle motion in *ideal gases*. By definition, we may say that an ideal gas is a gas that satisfies the empirically derived ideal gas law:

$$\boxed{nRT = pV,} \tag{2.46}$$

where $T$ is the absolute temperature of the gas (measured in $°K$), $n$ its substance in moles (as explained below), $V$ the volume that it occupies, $p$ the pressure that it exerts on its container, and

$$R = 8.314472 \ JK^{-1}mol^{-1}$$

is the gas constant.

Given this definition, our goal will be to derive certain essential properties of an ideal gas by analyzing, statistically, the behavior of the molecules or atoms that compose it. To this end we assume that an ideal gas with a total of $N$ particles is enclosed in a rectangular box $B = [0, a] \times [0, b] \times [0, c] \subset \mathbb{R}^3$. If the box is not unusually large, then gravitational effects can be disregarded and, by implication, the density of the gas can be assumed to be constant throughout the box and the atoms or molecules of which the gas consists can be assumed to be moving in all spatial directions with equal probability. Consequently, the pressure $p$ exerted on the bottom face $[0, a] \times [0, b] \times \{0\}$ is equal to the pressure exerted on each of the other five faces of the box, and it is this uniform pressure $p$ that equation (2.46) is referring to.

Since the pressure exerted on a surface segment by a perpendicular force is equal to the magnitude of the force divided by the area of the segment, it follows that we need to determine the perpendicular force that the gas particles exert on the bottom face by colliding with that bottom face. To do so, we observe that each microscopic collision of a particle with the bottom face effects a change of the particle's momentum and thereby induces a force because, according to Newton's second fundamental law of mechanics, force is equal to change in momentum over change in time. To determine the change in momentum in question, we will further assume that the temperature of the gas is the same as the temperature of the box. For in this case, there will be no net-transfer of energy between the gas and the box, and by implication, the collision of a gas particle with the bottom face may be assumed to simply reverse the sign of the $z$-component of the particle's momentum vector. Denoting the components of this momentum vector by $mu$, $mv$, and $mw$ respectively (where $m$ is the particle's mass), it follows that the momentum after the collision is

$$m \begin{pmatrix} u \\ v \\ -w \end{pmatrix}$$

and the corresponding change in momentum is

$$m \left( \begin{pmatrix} u \\ v \\ -w \end{pmatrix} - \begin{pmatrix} u \\ v \\ w \end{pmatrix} \right) = m \begin{pmatrix} 0 \\ 0 \\ -2w \end{pmatrix}.$$

Since this change-in-momentum vector is evidently perpendicular to the bottom face $[0, a] \times [0, b] \times \{0\}$, we can take it as it is and do not need to extract its vertical component.

To proceed, we need to determine the number of particles that collide with the bottom face in a small amount of time $\Delta t$ and then use this number to calculate the corresponding total change in momentum. Since a particle with velocity $w$ in the $z$-direction changes its $z$-coordinate by the amount $w\Delta t$ in the time span $\Delta t$, we may infer that only those particles will collide with the bottom face whose initial $z$-coordinate—at the beginning of the time interval—satisfies the inequality

$$w\Delta t \leq -z$$

or, equivalently,

$$w \leq -\frac{z}{\Delta t}. \tag{2.47}$$

If we consider the so-called *phase space* of all possible position-momentum points $(x, y, z, u, v, w)$ to be our sample space $S$, then both $z$ and $w$ may be regarded as random variables on $S$ (that is to say, for a given point $s = (x, y, z, u, v, w) \in S$ we set $z(s) := z$ and $w(s) := w$). Furthermore, $z$ and $w$ are independent because there is no physical reason to assume that the value of the velocity $w$ should be in any way dependent upon the position coordinate $z$. Consequently, the joint density of $z$ and $w$ on $\mathbb{R}^2$ is the product of the marginal densities of $z$ and $w$. So as we denote by $f(w)$ the density of $w$ and as we observe that $z$ must be uniformly distributed over $[0, c]$ (because the density of the gas is the same throughout the box), we may infer that the joint density of $z$ and $w$ is equal to

$$\begin{cases} f(w)/c & \text{if } (z, w) \in [0, c] \times \mathbb{R}, \\ 0 & \text{otherwise.} \end{cases}$$

Thus the probability for a particle to satisfy condition (2.47) is

$$\int_0^c \int_{-\infty}^{-z/\Delta t} \frac{f(w)}{c} \, dw \, dz,$$

and, by implication, the corresponding total number of particles (that satisfy this condition) is

$$N \int_0^c \int_{-\infty}^{-z/\Delta t} \frac{f(w)}{c} \, dw \, dz.$$

Since each of these particles changes its momentum (in the vertical direction) by $-2mw$ (due to its colliding with the bottom face), we may infer that the total change in momentum over the change in time $\Delta t$ is

$$-\frac{2mN}{c\Delta t} \int_0^c \int_{-\infty}^{-z/\Delta t} w f(w) \, dw \, dz.$$

Consequently, the pressure on the bottom face is

$$p = -\lim_{\Delta t \to 0} \frac{2mN}{abc\Delta t} \int_0^c \int_{-\infty}^{-z/\Delta t} w f(w) \, dw \, dz. \tag{2.48}$$

Since at this point the density $f$ is still unknown, we cannot evaluate the inner integral $\int_{-\infty}^{-z/\Delta t} w f(w) \, dw$. But what we can do is to carry out the integration with respect to $z$ because the integrand $w f(w)$ does not depend on $z$. Hence we need to change the order of integration, and to this end it is helpful to take

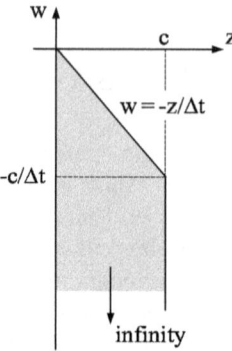

Figure 2.15: the region of integration.

a look at the region of integration in the $zw$-plane that the double integral in equation (2.48) specifies. For given the picture in Figure 2.15, it is easy to see that in order to integrate $wf(w)$ over this region with respect to $z$ first and $w$ second, we need to break this region up into the infinite rectangle $[0, c] \times (-\infty, -c/\Delta t]$ and the triangle with vertices at $(0,0)$, $(0, -c/\Delta t)$, and $(c, -c/\Delta t)$. This yields

$$
p = -\lim_{\Delta t \to 0} \frac{2mN}{abc\Delta t} \int_{-c/\Delta t}^{0} \int_{0}^{-w\Delta t} wf(w) \, dz \, dw
$$

$$
-\lim_{\Delta t \to 0} \frac{2mN}{abc\Delta t} \int_{-\infty}^{-c/\Delta t} \int_{0}^{c} wf(w) \, dz \, dw
$$

$$
= \lim_{\Delta t \to 0} \frac{2mN}{abc} \int_{-c/\Delta t}^{0} w^2 f(w) \, dw - \lim_{\Delta t \to 0^+} \frac{2mNc}{abc\Delta t} \int_{-\infty}^{-c/\Delta t} wf(w) \, dw
$$

$$
= \frac{2mN}{abc} \int_{-\infty}^{0} w^2 f(w) \, dw + \frac{2mN}{abc} \lim_{y \to -\infty} y \int_{-\infty}^{y} wf(w) \, dw.
$$

Since the (implicit) assumption that the integral $\int_{-\infty}^{0} wf(w) \, dw$ is convergent implies that

$$
\lim_{y \to -\infty} \int_{-\infty}^{y} wf(w) \, dw = 0,
$$

we may apply L'Hôpital's rule in conjunction with the first fundamental theorem of calculus to infer that

$$
\lim_{y \to -\infty} y \int_{-\infty}^{y} wf(w) \, dw = \lim_{y \to -\infty} \frac{\int_{-\infty}^{y} wf(w) \, dw}{1/y} = \lim_{y \to -\infty} \frac{yf(y)}{-1/y^2}
$$

$$
= -\lim_{y \to -\infty} y^3 f(y).
$$

Given that very large particle speeds in a gas are naturally very rare, we may further assume that the decrease of $f(y)$ as $y$ tends to $-\infty$ is sufficiently rapid to guarantee that the last of the limits above is equal to zero. (In fact, the decrease is exponential because $f$ is a normal density function as we shall see below.) Hence

$$p = \frac{2mN}{abc} \int_{-\infty}^{0} w^2 f(w)\, dw.$$

Noting further that the distribution of velocities may be assumed to be the same in all spatial directions, it follows that $f(w) = f(-w)$ and that, by implication,

$$\int_{-\infty}^{0} w^2 f(w)\, dw = \int_{0}^{\infty} w^2 f(w)\, dw.$$

Thus

$$p = \frac{mN}{abc} \int_{-\infty}^{\infty} w^2 f(w)\, dw = \frac{mN}{abc} E(w^2).$$

Using again the spatial uniformity of the distribution of velocities, we may infer that $E(u^2) = E(v^2) = E(w^2)$, and therefore,

$$p = \frac{mN}{3abc} E(u^2 + v^2 + w^2).$$

Denoting the volume $abc$ of the box by $V$ and observing that the kinetic energy of a gas particle is $E_{kin} = m(u^2 + v^2 + w^2)/2$, we find that

$$pV = \frac{2N}{3} E(E_{kin}). \qquad (2.49)$$

Since $n$ in (2.46) is related to $N$ in (2.49) via the equation

$$n = \frac{N}{N_A},$$

where $N_A = 6.022142 \cdot 10^{23}\ mol^{-1}$ is Avogadro's constant (which equals the number of atoms in 12 grams of carbon-12), we may introduce Boltzmann's constant $k := R/N_A$ and combine (2.49) with (2.46) to arrive at the following conclusion:

$$\boxed{E(E_{kin}) = \frac{3kT}{2}.}$$

In other words, *the temperature of an ideal gas is a measure for the expected kinetic energy of the particles that compose the gas.*

Our next objective is to derive an explicit formula for the density $f(w)$ as well as for the joint density of the velocity vector $(u, v, w)$. To do so, we denote by $g(u, v, w)\, du\, dv\, dw$ the infinitesimal probability for $(u, v, w)$ to be found in the infinitesimal rectangular box $[u, u + du] \times [v, v + dv] \times [w, w + dw]$. Since no spatial direction is a preferred direction as far as the motion of particles in an ideal gas is concerned, it follows that there exists a function $h$ such that

$$g(u, v, w) = h(u^2 + v^2 + w^2).$$

Furthermore, since the values of $u$, $v$, and $w$ may be assumed to be independent of each other, we may infer that

$$g(u, v, w) = f(u)f(v)f(w),$$

and therefore,

$$f(u)f(v)f(w) = h(u^2 + v^2 + w^2).$$

Differentiating both sides with respect to each of the three variables $u$, $v$, and $w$ yields

$$f'(u)f(v)f(w) = 2uh'(u^2 + v^2 + w^2),$$
$$f(u)f'(v)f(w) = 2vh'(u^2 + v^2 + w^2),$$
$$f(u)f(v)f'(w) = 2wh'(u^2 + v^2 + w^2),$$

and, by implication,

$$\frac{f'(u)f(v)f(w)}{2u} = \frac{f(u)f'(v)f(w)}{2v} = \frac{f(u)f(v)f'(w)}{2w} = h'(u^2 + v^2 + w^2).$$

As we divide these resulting equations by $f(u)f(v)f(w)$, we find that

$$\frac{f'(u)}{2uf(u)} = \frac{f'(v)}{2vf(v)} = \frac{f'(w)}{2wf(w)},$$

and using the same reasoning as in the derivation of (2.25) and (2.26) from (2.24), it follows that there are constants $C, D \in \mathbb{R}$ such that

$$f(w) = De^{Cw^2}.$$

Hence $g$ is a normal density function on $\mathbb{R}^3$ because

$$g(u, v, w) = f(u)f(v)f(w) = D^3 e^{Cu^2} e^{Cv^2} e^{Cw^2} = D^3 e^{C(u^2 + v^2 + w^2)}.$$

Thus, in setting

$$C := -\frac{1}{2\sigma^2},$$

we may infer that

$$D = \frac{1}{\sigma\sqrt{2\pi}},$$

and therefore,

$$g(u, v, w) = \frac{e^{-(u^2+v^2+w^2)/(2\sigma^2)}}{\sigma^3 \sqrt{2\pi}^3}.$$

In order to determine $\sigma$, we make use of the fact that

$$f(w) = \frac{e^{-w^2/(2\sigma^2)}}{\sigma\sqrt{2\pi}}$$

is a normal density with mean $E(w) = 0$. For in the light of this fact, it follows that

$$\sigma^2 = E(w^2) - E(w)^2 = E(w^2) = \frac{2}{3m}E(E_{kin}) = \frac{kT}{m},$$

and therefore,

$$g(u, v, w) = \frac{\sqrt{m}^3 e^{-m(u^2+v^2+w^2)/(2kT)}}{\sqrt{2\pi kT}^3}.$$

# Exercises

**2.10.1.** Assume that an ideal gas, consisting of particles of mass $m$, is kept at a constant temperature $T$.

a) What is the expected value of the $z$-velocity component (which we denoted by $w$) of a particle in the gas and what is the expected value of the $z$-direction speed?

b) What is the expected value of the square of a particle's speed (in three-dimensional space)?

c) What is the expected value of a particle's speed? (Be careful: the answer is not just the square root of the answer to b). It's a bit more involved.)

d) What is the expected numerical value of the speed of a hydrogen atom in a hydrogen gas at room temperature $(20°C = 293.15°K)$? Note: the mass of a hydrogen atom is $(0.00100794/N_A)\,kg$

e) What is the standard deviation of the speed of a hydrogen atom in a hydrogen gas at room temperature $(20°C)$?

**2.10.2.** Assume that $f : \mathbb{R} \to \mathbb{R}$ is the density of a random variable $X : S \to \mathbb{R}$. Show that $E(H(X)) = 0$ if $f(-x) = f(x)$ and $H(-x) = -H(x)$ for all $x \in \mathbb{R}$ (and if $\int_0^\infty H(x)f(x)\,dx$ is convergent).

# 2.11   Climate Change and Correlation

One of the most prominent topics in current-day scientific research is the question of whether and to what extent the earth's climate is affected by the increase in atmospheric $CO_2$ that is caused by the burning of fossil fuels in modern industrial societies. Naturally, this issue is highly complex and cannot be addressed with confidence by anyone who doesn't have extensive knowledge of the relevant theories and data. In fact, in posing the question, as we just did, as an actual question, we already are, in a sense, overstepping the narrow bounds that any casual discourse on climate science had better be confined to. For according to what seems to be the mainstream point of view, this "question of whether and to what extent" has been settled conclusively: the increase in $CO_2$ does and will cause the earth's temperature to rise and to rise dramatically—in all likelihood—if nothing is done to limit emissions.

There are of course detractors—and even some with impeccable credentials—but supposedly their arguments are untenable and their numbers minute. The detractors, though, do not agree. According to them, the mainstream view is poorly supported by the data at best and downright fraudulent at worst. Moreover, their seeming isolation, so they say, is not actual but really merely seeming. For in an unwholesome climate of persecution and censorship the numerous dissenters that are there are loath to speak up lest their careers be totally ruined.

It goes without saying that there is no hope for us in this present exposition to sort these matters out. There appears to be a near universal consensus that $CO_2$ is a greenhouse gas that does cause some warming. But the problem of whether recent or not so recent warming or cooling trends are in fact what they appear to be and are indeed caused by whatever causal agents people have identified is a matter that probably is best left to those who claim to have the requisite insight—rightly or wrongly.

However, what we surely *can* do is to try to take a strictly neutral look at some of the relevant issues and then discuss some mathematical methods pertaining to them. The basic claim, as we have said, is that the earth's temperature rises as the concentration of atmospheric $CO_2$ increases. One way to support this claim is to examine the earth's historical temperature-versus-$CO_2$ record. As it turns out, what researchers have found in this regard is a very close *correlation*: when $CO_2$ goes up the temperature goes up as well, and when $CO_2$ goes down so does the temperature. But unfortunately, correlation is not the same as causation, and hence the detractors have argued that this observed historical synchrony is very much misleading. For it is always—or most of the time—the temperature that changes first and the $CO_2$ concentration that follows. Surprisingly, there even is a reason for this lagging behind of $CO_2$ that all sides seem to accept. For it is a well established scientific fact that colder water can absorb more $CO_2$ than water that is warmer. That is to say, as the atmospheric temperature rises—for whatever reasons—the oceans become warmer as well and thus release some $CO_2$ in the process. Furthermore, since the oceans are vast and since water has a much greater capacity for storing heat than air, so the detractors point out, the ocean temperature lags behind. The atmosphere gets warmer, but the oceans take a few centuries to really catch up, and therefore, the increase in $CO_2$ is bound to be delayed.

In summary this argument shows, or so it may seem, that the causal culprit is the temperature and not $CO_2$. But here the detractors are going too far, or so we are told by their mainstream opponents. For while it is true that warmer oceans absorb less $CO_2$ than colder ones, it also is true undoubtedly that $CO_2$ is a greenhouse gas. So the warming of the weather causes the $CO_2$ concentration to rise by reducing ocean-absorption, but this rise in $CO_2$ in turn causes the warming to further accelerate. So there is a feedback effect that cannot be neglected, and neglecting the potential dangers of fossil fuel burning would therefore be utterly reckless—supposedly or actually, the jury is still out.

There are numerous arguments of this sort that people employ to support or defeat the standard climate teachings. But let's not get bogged down in them, and let's examine instead some hypothetical graphs that show some lagging correlations. In Figure 2.16 on the left we see a $CO_2$ graph that lags behind the parallel temperature graph by a fairly small margin, in the middle diagram the margin is larger, and on the right it is so large that the correlation is almost inverted. That is to say, in the third diagram on the right an increase in one graph corresponds to a decrease in the other and vice versa.

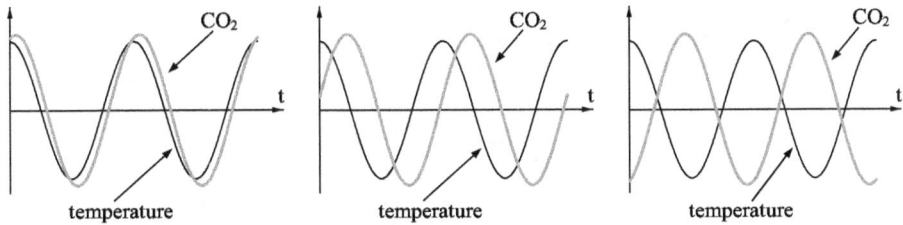

Figure 2.16: hypothetical $CO_2$-versus-temperature graphs.

In order to approach the question of how the differences in correlation that these diagrams display can be quantified, we notice to begin with that the product of the functions that represent the temperature and $CO_2$ graphs, respectively, is mostly positive on the left, mostly negative on the right and fluctuating between negative and positive values in the middle. This is so because on the left the two functions are positive and negative in synchrony, on the right one function is positive where the other is negative and vice versa, and in the middle there are intervals where the two functions have the same sign and other intervals where their signs are opposite. Consequently, the integral of the product is strongly positive on the left, strongly negative on the right, and relatively small in absolute value in the middle.

So perhaps then, the correlation between two graphs should simply be measured by the integral of their product. Let's see what this yields for each of the pairs of graphs in Figure 2.16: the functions that describe these graphs we chose to be, for simplicity,

$$X(t) := \cos(2t) \quad \text{(temperature graph in all three diagrams)}$$

and

$$Y_1(t) := 1.1 \cos(2t - 0.3) \quad (CO_2 \text{ graph on the left}),$$
$$Y_2(t) := 1.1 \cos(2t - 1.4) \quad (CO_2 \text{ graph in the middle}),$$
$$Y_3(t) := 1.1 \cos(2t - 2.8) \quad (CO_2 \text{ graph on the right}),$$

where $t \in [0, 2\pi]$. Using a computer or calculating by hand, we readily find that

$$\int_0^{2\pi} X(t)Y_1(t)\,dt = 1.1\pi \cos(0.3) \approx 3.301,$$

$$\int_0^{2\pi} X(t)Y_2(t)\,dt = 1.1\pi \cos(1.4) \approx 0.587, \tag{2.50}$$

$$\int_0^{2\pi} X(t)Y_3(t)\,dt = 1.1\pi \cos(2.8) \approx -3.256.$$

(Note: the scales on the horizontal and vertical axes in Figure 2.16 are not equal and the graphs are therefore distorted, but that's of course irrelevant as far as the values of the integrals above are concerned.)

Given that we chose to use the letters '$X$' and '$Y$' to denote the functions in Figure 2.16, there seems to be a suggestion that these functions are somehow to be viewed as random variables. But what exactly is random about them and what is the sample space on which they are defined? Well, first of all, an actual temperature or $CO_2$ graph doesn't look nearly as smooth and regular as the cosine graphs in Figure 2.16. So there is a natural randomness about temperature data that makes it reasonable for us to ask how likely it is to measure a value that is contained in a given temperature interval. And the way to determine this likelihood is naturally to consult the historical precedent. That is to say, the likelihood or probability here in question can be considered to be the relative amount of time during which historical temperature values were contained in that given interval. Since the overall historical time range in the diagrams in Figure 2.16 happens to be the interval $[0, 2\pi]$ on the $t$-axis, it makes sense to regard this interval—equipped with the uniform density—to be the sample space on which $X$, $Y_1$, $Y_2$, and $Y_3$ are defined. Thus it seems appropriate to replace the integrals in (2.50) with the following expected values:

$$E(XY_1) = \frac{1.1\pi \cos(0.3)}{2\pi} \approx 0.525,$$

$$E(XY_2) = \frac{1.1\pi \cos(1.4)}{2\pi} \approx 0.093, \qquad (2.51)$$

$$E(XY_3) = \frac{1.1\pi \cos(2.8)}{2\pi} \approx -0.518.$$

But is an expected value of the product of two random variables really a good measure for how strongly these variables are correlated? What for example would happen if we replaced the random variables $X(t)$ and $Y_1(t)$ in the diagram on the left in Figure 2.16 by, say, $X(t) - 2$ and $Y_1(t) + 3$? In that case the graphs would be shifted relative to each other in the vertical direction, but the correlation between them would basically remain unchanged—they would still increase and decrease in almost perfect synchrony by exactly the same amounts as before. However, the expected value of their product would be drastically different:

$$E((X-2)(Y_1+3)) = \int_0^{2\pi} (\cos(2t) - 2)(1.1\cos(2t - 0.3) + 3)\frac{dt}{2\pi}$$
$$\approx -5.475.$$

In order to guard against this possibility of arbitrarily distorting expected values by adding constants (like $-2$ and $3$), we need to refer the values of a random variable back to their mean. That is to say, we need to subtract the mean from the random variable, and thus we are led to introduce the following definition:

**2.11.1 Definition.** The *covariance* of two random variables $X, Y : S \to \mathbb{R}$ is

$$Cov(X, Y) := E((X - E(X))(Y - E(Y))).$$

Note: this definition trivially implies that $Cov(X, X) = Var(X)$.

**2.11.2 Example.** The mean values of the random variables $X - 2$ and $Y_1 + 3$ are

$$E(X - 2) = \int_0^{2\pi} (\cos(2t) - 2)\frac{dt}{2\pi} = -2$$

and

$$E(Y_1 + 3) = \int_0^{2\pi} (1.1\cos(2t - 0.3) + 3)\frac{dt}{2\pi} = 3,$$

respectively. Hence

$$Cov(X - 2, Y_1 + 3) = E((X - 2 - E(X - 2))(Y_1 + 3 - E(Y_1 + 3)))$$
$$= E((X - 2 - (-2))(Y_1 + 3 - 3))$$
$$= E(XY_1) \approx 0.525.$$

So due to the subtraction of the mean values in the definition of the covariance we find here again the same value as in (2.51).

Inspired by this example, we formulate the following theorem:

**2.11.3 Theorem.** *If $X$ and $Y$ are random variables on a sample space $S$ and if $c$ and $d$ are given real numbers, then*

$$\mathrm{Cov}(X + c, Y + d) = \mathrm{Cov}(X, Y).$$

*Proof.* Using Theorems 2.3.4 and 2.3.5, we find that

$$\begin{aligned}
\mathrm{Cov}(X + c, Y + d) &= E((X + c - E(X + c))(Y + d - E(Y + d))) \\
&= E((X + c - E(X) - c)(Y + d - E(Y) + d)) \\
&= E((X - E(X))(Y - E(Y))) = \mathrm{Cov}(X, Y),
\end{aligned}$$

as desired. $\qquad\square$

Another useful result concerning covariances is this:

**2.11.4 Theorem.** *If $X, Y : S \to \mathbb{R}$ are given random variables, then*

$$\mathrm{Cov}(X, Y) = E(XY) - E(X)E(Y).$$

*Proof.* Using again Theorems 2.3.4 and 2.3.5, we may infer that

$$\begin{aligned}
\mathrm{Cov}(X, Y) &= E((X - E(X))(Y - E(Y))) \\
&= E(XY - E(X)Y - E(Y)X + E(X)E(Y)) \\
&= E(XY) - E(X)E(Y) - E(Y)E(X) + E(X)E(Y) \\
&= E(XY) - E(X)E(Y),
\end{aligned}$$

as desired. $\qquad\square$

**2.11.5 Corollary.** *If $X, Y : S \to \mathbb{R}$ are independent random variables, then* $\mathrm{Cov}(X, Y) = 0$.

*Proof.* The proof is a trivial consequence of Theorems 2.9.8 and 2.11.4. $\qquad\square$

The statement of Corollary 2.11.5 makes perfect sense, because if $X$ and $Y$ are independent, then, presumably, knowledge of the values of $X$ is independent of knowledge of the values of $Y$, and thus there is no correlation. So if indeed the covariance is a measure of how much the values of $X$ and $Y$ are correlated with each other, then we ought to expect that the covariance vanishes in the case where $X$ and $Y$ are independent.

However, the question of whether the covariance is indeed an adequate measure of correlation hasn't really been settled yet, and, in fact, the answer

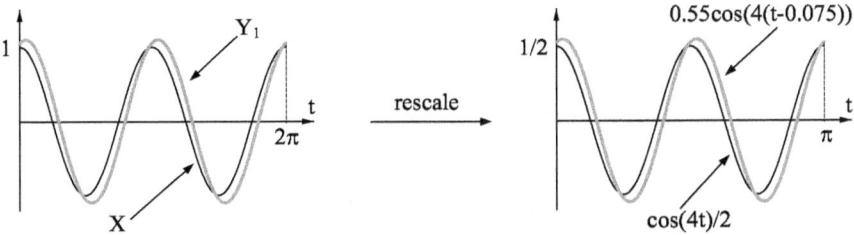

Figure 2.17: rescaling the graphs of $X$ and $Y_1$.

is "no." For suppose that we rescale the diagram on the left in Figure 2.16 by a factor two by replacing the sample space $[0, 2\pi]$ with $[0, \pi]$, $X(t) = \cos(2t)$ with $\cos(4t)/2$, and $Y_1(t) = 1.1\cos(2(t - 0.15))$ with $0.55\cos(4(t - 0.075))$ (Figure 2.17). In that case the labels on the axes change but otherwise the picture remains completely invariant. Consequently, any reasonable quantitative measure of correlation ought to remain invariant as well. Unfortunately, though, the covariance of the rescaled random variables is not the same as the covariance of $X$ and $Y_1$:

$$E(\cos(4t)/2) = \int_0^\pi \frac{\cos(4t)}{2} \frac{dt}{\pi} = 0,$$

$$E(0.55\cos(4(t - 0.075))) = \int_0^\pi 0.55\cos(4(t - 0.075)) \frac{dt}{\pi} = 0,$$

and therefore,

$$\begin{aligned} \mathrm{Cov}\left(\frac{\cos(4t)}{2}, 0.55\cos(4(t - 0.075))\right) &= \int_0^\pi \frac{0.55\cos(4t)\cos(4(t - 0.075))}{2} \frac{dt}{\pi} \\ &= \int_0^{2\pi} \frac{1.1\cos(2u)\cos(2u - 0.3))}{4} \frac{du}{2\pi} \\ &= \frac{E(XY_1)}{4}. \end{aligned}$$

So the rescaling reduces the covariance to a quarter of its original value. And this is not surprising because the defining equation for the covariance in Definition 2.11.1 is evidently not invariant under changes in the absolute magnitudes of $X - E(X)$ and $Y - E(Y)$. But what are these absolute magnitudes in the context of probability theory? They are the standard deviations $\sigma(X) = \sqrt{\mathrm{Var}(X)}$ and $\sigma(Y) = \sqrt{\mathrm{Var}(Y)}$, respectively. By implication, it is by the product of these two standard deviations by which we need to divide the covariance in order to achieve invariance under rescalings.

**2.11.6 Definition.** The *correlation coefficient* of two random variables $X, Y$ :

$S \to \mathbb{R}$ with positive standard deviations (i.e., $\sigma(X), \sigma(Y) > 0$) is

$$\rho(X, Y) := \frac{\mathrm{Cov}(X, Y)}{\sigma(X)\sigma(Y)}.$$

*Remark.* The properties stated in Theorem 2.11.3 and Corollary 2.11.5 are valid for the correlation coefficient as well. That is to say,

$$\boxed{\rho(X + c, Y + d) = \rho(X, Y)} \qquad (2.52)$$

for all $c, d \in \mathbb{R}$ by Theorems 2.3.7c and 2.11.3, and if $X$ and $Y$ are independent, then $\rho(X, Y) = 0$ by Corollary 2.11.5.

**2.11.7 Example.** Considering again the random variables $X(t) = \cos(2t)$ and $Y_1(t) = 1.1\cos(2t - 0.3)$ on $[0, 2\pi]$, we find that $E(X) = E(Y_1) = 0$, and therefore,

$$\mathrm{Var}(X) = E(X^2) = \int_0^{2\pi} \cos^2(2t)\, \frac{dt}{2\pi} = 0.5$$

and

$$\mathrm{Var}(Y_1) = E(Y_1^2) = \int_0^{2\pi} 1.1^2 \cos^2(2t - 0.3)\, \frac{dt}{2\pi} = 0.605.$$

Hence

$$\rho(X, Y_1) = \frac{\mathrm{Cov}(X, Y_1)}{\sigma(X)\sigma(Y_1)} = \frac{E(XY_1)}{\sqrt{\mathrm{Var}(X)\,\mathrm{Var}(Y_1)}} \approx \frac{0.525}{\sqrt{0.5 \cdot 0.605}} \approx 0.955,$$

and similarly, we also see that

$$\rho(X, Y_2) \approx \frac{0.093}{\sqrt{0.5 \cdot 0.605}} \approx 0.170$$

and

$$\rho(X, Y_3) \approx \frac{-0.518}{\sqrt{0.5 \cdot 0.605}} \approx -0.942.$$

Furthermore, for the rescaled random variables in Figure 2.17 we find that

$$\mathrm{Var}(\cos(4t)/2) = \int_0^\pi \frac{\cos^2(4t)}{4}\, \frac{dt}{\pi} = \int_0^{2\pi} \frac{\cos^2(2u)}{4}\, \frac{du}{2\pi} = \frac{\mathrm{Var}(X)}{4}$$

and

$$\mathrm{Var}(0.55\cos(4(t - 0.075))) = \int_0^\pi 0.55^2 \cos^2(4t - 0.3)\, \frac{dt}{\pi}$$
$$= \int_0^{2\pi} \frac{1.1^2 \cos^2(2u - 0.3)}{4}\, \frac{du}{2\pi} = \frac{\mathrm{Var}(Y_1)}{4},$$

and therefore,

$$\rho(\cos(4t)/2, 0.55\cos(4(t-0.075)))) = \frac{E(XY_1)/4}{\sqrt{\mathrm{Var}(X)\,\mathrm{Var}(Y_1)/16}} = \rho(X, Y_1).$$

In other words, the rescaling does indeed leave the correlation coefficient invariant, and this observation is also confirmed by the following theorem:

**2.11.8 Theorem.** *If $X, Y : S \to \mathbb{R}$ are random variables with positive standard deviations and if $\lambda$ and $\mu$ are non-zero real numbers, then*

$$\rho(\lambda X, \mu Y) = \frac{\lambda\mu}{|\lambda\mu|}\rho(X, Y) = \pm\rho(X, Y).$$

*In particular,*

$$\rho(\lambda X, \mu Y) = \rho(X, Y),$$

*whenever $\lambda$ and $\mu$ are both positive or both negative, and*

$$\rho(\lambda X, \mu Y) = -\rho(X, Y),$$

*whenever the signs of $\lambda$ and $\mu$ are opposite.*

*Proof.* Using Theorems 2.11.4, 2.3.5b, and 2.3.7b, we find that

$$\rho(\lambda X, \mu Y) = \frac{E(\lambda\mu XY) - E(\lambda X)E(\mu Y)}{\sqrt{\mathrm{Var}(\lambda X)\,\mathrm{Var}(\mu Y)}} = \frac{\lambda\mu(E(XY) - E(X)E(Y))}{\sqrt{\lambda^2\mu^2\,\mathrm{Var}(X)\,\mathrm{Var}(Y)}}$$

$$= \frac{\lambda\mu}{|\lambda\mu|}\rho(X, Y),$$

and this equation immediately implies the statement of the theorem. □

The three correlation coefficients that we computed in Example 2.11.7 were all between $-1$ and $1$. Theorem 2.11.10 below shows that this was not a coincidence.

**2.11.9 Lemma.** *If $Z : S \to \mathbb{R}$ is a random variable such that $E(Z^2) = 0$, then*

$$P(Z = 0) = P(\{s \in S \mid Z(s) = 0\}) = 1.$$

*Proof.* We will assume that $S$ is continuous as the discrete case is completely analogous. Setting $M := \{s \in S \mid Z(s) = 0\}$, it is sufficient to show that the probability of $N := S \setminus M$ is zero. If it were the case that $P(N) > 0$, then it would also be the case that

$$E(Z^2) = \int_S Z(s)^2 p(s)\,ds \geq \int_N Z(s)^2 p(s)\,ds > 0$$

because $Z(s)^2 > 0$ for all $s \in N$. Since this contradicts the assumption $E(Z^2) = 0$, we may infer that $P(N) = 0$, as desired. □

**2.11.10 Theorem.** *If $X, Y : S \to \mathbb{R}$ are random variables with positive standard deviations, then*

$$-1 \leq \rho(X, Y) \leq 1.$$

*Furthermore, $\rho(X, Y) = \pm 1$ if and only if there are constants $m, b \in \mathbb{R}$ and a set $M \subset S$ such that $m \neq 0$, $P(M) = 1$, and $Y(s) = mX(s) + b$ for all $s \in M$. More precisely, in this latter case we have $\rho(X, Y) = 1$ if $m > 0$ and $\rho(X, Y) = -1$ if $m < 0$. (Note: the condition $P(M) = 1$ does not imply that $M = S$ because if, for example, $S = [0, 1]$ and $M = (0, 1]$ and if the density on $S$ is uniform, then $P(M) = 1$ but $M \neq S$.)*

*Proof.* Let $U, V : S \to \mathbb{R}$ be random variables such that $E(U^2) > 0$, and let $\lambda \in \mathbb{R}$. Since $(\lambda U(s) + V(s))^2 \geq 0$ for all $s \in S$, it follows that

$$E((\lambda U + V)^2) \geq 0.$$

Hence

$$0 \leq E(\lambda^2 U^2 + 2\lambda UV + V^2) = \lambda^2 E(U^2) + 2\lambda E(UV) + E(V^2).$$

for all $\lambda \in \mathbb{R}$, and therefore, the polynomial

$$Q(x) := x^2 E(U^2) + 2x E(UV) + E(V^2)$$

has at most one root. Using the quadratic formula in conjunction with the assumption $E(U^2) > 0$, we may infer that the roots of $Q$ are

$$\frac{-E(UV) \pm \sqrt{E(UV)^2 - E(U^2)E(V^2)}}{E(U^2)}. \tag{2.53}$$

But since $Q$ has at most one root, the term under the square root must be less than or equal to zero, that is,

$$E(UV)^2 - E(U^2)E(V^2) \leq 0,$$

or equivalently,

$$|E(UV)| \leq \sqrt{E(U^2)E(V^2)}.$$

Substituting $X - E(X)$ and $Y - E(Y)$ for $U$ and $V$, respectively, it follows that

$$|\text{Cov}(X, Y)| = |E(X - E(X))E(Y - E(Y))| \leq \sigma(X)\sigma(Y),$$

and therefore, $|\rho(X, Y)| \leq 1$, as desired. Furthermore, if there is a set $M \subset S$ such that $P(M) = 1$ and $Y(s) = mX(s) + b$ for all $s \in M$ and some $m \neq 0$

and some $b \in \mathbb{R}$, then, according to (2.52) and Theorem 2.11.8 (which both remain applicable because $P(M) = 1$), we have

$$\rho(X, Y) = \rho(X, mX + b) = \rho(X, mX) = \frac{m}{|m|} \rho(X, X)$$

$$= \pm \frac{E((X - E(X))^2)}{\sigma(X)^2} = \pm \frac{\text{Var}(X)}{\text{Var}(X)} = \pm 1 = \frac{m}{|m|},$$

and in particular, $\rho(X, Y) = 1$ if $m > 0$ and $\rho(X, Y) = -1$ if $m < 0$. Finally and conversely, if $\rho(X, Y) = \pm 1$, then the resubstitution $U := X - E(X)$ and $V := Y - E(Y)$ yields

$$|E(UV)| = \sqrt{E(U^2)E(V^2)},$$

and, by implication, the root-term in (2.53) is zero. Hence $Q$ has exactly one root, namely $\lambda = -E(UV)/E(U^2)$. So for this value $\lambda$ it is the case that

$$0 = Q(\lambda) = \lambda^2 E(U^2) + 2\lambda E(UV) + E(V^2) = E((\lambda U + V)^2). \tag{2.54}$$

But according to Lemma 2.11.9 (with $Z := \lambda U + V$), this latter equation implies that $M := \{s \in S \mid \lambda U(s) + V(s) = 0\}$ is a set of full probability. That is to say, $P(M) = 1$ and

$$0 = \lambda U(s) + V(s) = \lambda X(s) - \lambda E(X) + Y(s) - E(y)$$

for all $s \in M$. So in setting $m := -\lambda$ and $b := \lambda E(X) + E(Y)$, we may infer that $Y(s) = mX(s) + b$ for all $s \in M$. Moreover, the assumption that $\sigma(Y)$ is positive implies that $m = -\lambda$ is different from zero because otherwise (2.54) would imply that $0 = E(V^2) = \text{Var}(Y)$. $\qquad\square$

Given the statement of Theorem 2.11.10 and given the results of Example 2.11.7, we are now justified in formulating the following general rules for any random variables $X, Y : S \to \mathbb{R}$ whose standard deviations are positive:

- There is a strong positive correlation between $X$ and $Y$ if and only if $\rho(X, Y) \approx 1$ (i.e., $Y \approx mX + b$ with $m > 0$).
- There is a strong negative or inverse correlation between $X$ and $Y$ if and only if $\rho(X, Y) \approx -1$ (i.e., $Y \approx mX + b$ with $m < 0$).
- The correlation between $X$ and $Y$ is weak if and only if $\rho(X, Y) \approx 0$.

In the light of the first of these rules, we should expect that the random variables $X(t) = \cos(2t)$ and $Y_1(t) = 1.1\cos(2t - 0.3)$ satisfy an approximate equation of the form $Y \approx mX + b$ with $m > 0$, because in Example 2.11.7 we found that the correlation coefficient $\rho(X, Y_1) \approx 0.955$ is very close to one. And indeed, as we apply the familiar addition law

$$\cos(\alpha - \beta) = \cos(\alpha)\cos(\beta) + \sin(\alpha)\sin(\beta)$$

to $Y_1$, we find that

$$Y_1(t) = 1.1\cos(0.3)\cos(2t) + 1.1\sin(0.3)\sin(2t)$$
$$\approx 1.051X(t) + 0.325\sin(2t),$$
(2.55)

and therefore,

$$1.051X(t) - 0.325 \le Y_1(t) \le 1.051X(t) + 0.325.$$

So the points $(X(t), Y_1(t)) \in \mathbb{R}^2$ will always be located between the lines that are described by the equations $y = 1.051x - 0.325$ and $y = 1.051x + 0.325$. To make this fact visually apparent, we can plot these lines together with the curve $c_1(t) = (X(t), Y_1(t))$ in an $xy$-coordinate system. The resulting image is shown in Figure 2.18. Moreover, as we plot as well the curves $c_2(t) = (X(t), Y_2(t))$

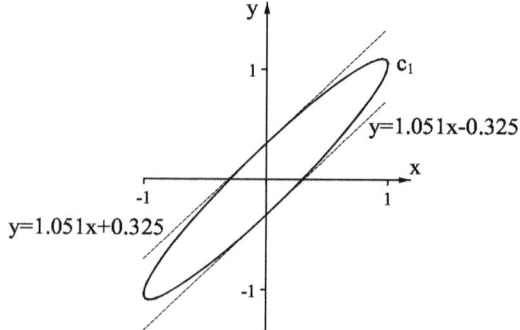

Figure 2.18: the curve $c_1 = (X, Y_1)$.

and $c_3(t) = (X(t), Y_3(t))$ (Figure 2.19), we readily notice that the narrowly elliptic curve $c_1$ stays relatively close to a line with positive slope $m \approx 1.051$, that $c_3$ stays close to a line with about the same slope but negative, and that $c_2$ does not really follow any line at all. This observation is nicely consistent with the fact that $\rho(X, Y_1) \approx 1$, $\rho(X, Y_3) \approx -1$, and $\rho(X, Y_2) \approx 0$.

**2.11.11 Example.** For a given $n \in \mathbb{N}$ we wish to compute the correlation coefficient of the random variables $X(s) := s$ and $Y_n(s) := \sin(ns)$ on the sample space $S = [0, 2\pi]$, where the density on $S$ is assumed to be uniform. Using a computer or calculating by hand, we find that

$$E(X) = \int_0^{2\pi} s\,\frac{ds}{2\pi} = \pi,$$

$$E(Y_n) = \int_0^{2\pi} \sin(ns)\,\frac{ds}{2\pi} = 0,$$

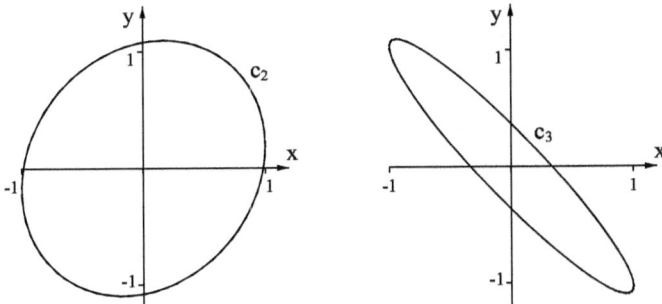

Figure 2.19: the curves $c_2 = (X, Y_2)$ and $c_3 = (X, Y_3)$.

$$E(X^2) = \int_0^{2\pi} s^2 \frac{ds}{2\pi} = \frac{4\pi^2}{3},$$

$$E(Y_n^2) = \int_0^{2\pi} \sin^2(ns) \frac{ds}{2\pi} = \frac{1}{2},$$

and

$$E(XY_n) = \int_0^{2\pi} s \sin(ns) \frac{ds}{2\pi} = -\frac{1}{n}.$$

Hence

$$\rho(X, Y_n) = \frac{E(XY_n) - E(X)E(Y_n)}{\sqrt{(E(X^2) - E(X)^2)(E(Y_n^2) - E(Y_n)^2)}} = -\frac{\sqrt{6}}{\pi n},$$

and therefore,

$$\lim_{n \to \infty} \rho(X, Y_n) = 0.$$

This result is not surprising because, as $n$ tends to infinity, the period of $Y_n$ shrinks to zero. That is to say, as $n$ increases, $Y_n$ oscillates ever more rapidly between $-1$ and $1$, and its correlation to the steadily increasing straight-line function $X$ therefore grows weaker accordingly.

**2.11.12 Example.** For a given $n \in \mathbb{N}$ we wish to compute the correlation coefficient of the random variables $X(s) := 2\sin(s)$ and $Y_n(s) := \sin(s - 1/n)$ on the sample space $S = [0, 2\pi]$, where the density on $S$ is again assumed to be uniform. This time we find that

$$E(X) = \int_0^{2\pi} 2\sin(s) \frac{ds}{2\pi} = 0,$$

$$E(Y_n) = \int_0^{2\pi} \sin(s - 1/n) \frac{ds}{2\pi} = 0,$$

$$E(X^2) = \int_0^{2\pi} 4\sin^2(s) \frac{ds}{2\pi} = 2,$$

$$E(Y_n^2) = \int_0^{2\pi} \sin^2(s - 1/n) \frac{ds}{2\pi} = \frac{1}{2},$$

and

$$E(XY_n) = \int_0^{2\pi} 2\sin(s)\sin(s - 1/n) \frac{ds}{2\pi} = \cos(1/n).$$

Hence

$$\rho(X, Y_n) = \frac{E(XY_n) - E(X)E(Y_n)}{\sqrt{(E(X^2) - E(X)^2)(E(Y_n^2) - E(Y_n)^2)}} = \cos(1/n),$$

and therefore,

$$\lim_{n\to\infty} \rho(X, Y_n) = 1.$$

Again this result is not surprising because, as $n$ tends to infinity, the functions $Y_n$ converge to $Y(s) = \sin(s) = X(s)/2$, and because, according to Theorem 2.11.10, it is the case that $\rho(X, X/2) = 1$.

In all the examples that we discussed in this section so far, we computed covariances—and correlation coefficients thereby—by means of integrals over a sample space. But as a matter of course, if the joint density $f(x, y)$ of $X$ and $Y$ is known, we can also find the covariance $E(XY) - E(X)E(Y)$ by using either (2.40) or (2.42).

**2.11.13 Example.** In Example 2.8.10 we found the joint density of the random variables $X(s, t) = s$ and $Y(s, t) = st$ on $S = (0, 1] \times (0, 1]$ to be

$$f(x, y) = \begin{cases} 1/x & \text{if } 0 < y \le x \le 1, \\ 0 & \text{otherwise.} \end{cases}$$

Using this formula in conjunction with (2.40), we may infer that

$$E(X) = \int_{-\infty}^{\infty}\int_{-\infty}^{\infty} x f(x, y)\, dy\, dx = \int_0^1 \int_0^x \frac{x}{x}\, dy\, dx = \frac{1}{2},$$

$$E(Y) = \int_0^1 \int_0^x \frac{y}{x}\, dy\, dx = \frac{1}{4},$$

$$E(X^2) = \int_0^1 \int_0^x \frac{x^2}{x}\, dy\, dx = \frac{1}{3},$$

$$E(Y^2) = \int_0^1 \int_0^x \frac{y^2}{x}\, dy\, dx = \frac{1}{9},$$

and

$$E(XY) = \int_0^1 \int_0^x \frac{xy}{x} \, dy \, dx = \frac{1}{6}.$$

Hence

$$\rho(X, Y) = \frac{1/6 - (1/2) \cdot (1/4)}{\sqrt{1/3 - (1/2)^2}\sqrt{1/9 - (1/4)^2}} = \frac{\sqrt{3}}{\sqrt{7}} \approx 0.655.$$

**2.11.14 Example.** Referring back to the discrete random variables $X$ and $Y$ in Example 2.9.7, we find that

$$E(X) = 1 \cdot f_X(1) + 0 \cdot f_X(0) = 0.52,$$
$$E(Y) = 1 \cdot f_Y(1) + 0 \cdot f_Y(0) = 0.49,$$
$$E(X^2) = 1^2 \cdot f_X(1) + 0^2 \cdot f_X(0) = E(X) = 0.52,$$
$$E(Y^2) = E(Y) = 0.49,$$
$$E(XY) = 1 \cdot f(1,1) + 0 \cdot (f(1,0) + f(0,1) + f(0,0)) = 0.28$$

and, by implication,

$$\rho(X, Y) = \frac{0.28 - 0.52 \cdot 0.49}{\sqrt{0.52 - 0.52^2}\sqrt{0.49 - 0.49^2}} \approx 0.1009.$$

Consequently and as expected, the correlation between the female gender and the Democratic party affiliation is indeed slightly positive.

## Exercises

**2.11.15.** Use integrals over the sample space $S = (0, 1] \times (0, 1]$ to verify the result of Example 2.11.13.

**2.11.16.** Find the correlation coefficient of the random variables $X(s) = s$ and $Y(s) = s^2$ on each of the following sample spaces:

a) $S = [0, 1]$ with uniform density,

b) $S = [-1, 1]$ with uniform density,

c) $S = [0, 1]$ with $P(E) = \int_E 3s^2 \, ds$ for all events $E \subset S$.

**2.11.17.** Find the correlation coefficient for each of the pairs of random variables given below and find the limit of this coefficient as $n$ tends to infinity. The sample space in each case is $S = [0, 1]$ with uniform density, and $n$ is an arbitrary positive integer.

a) $X_n(s) := s^n$, $Y_n(s) := s^{n+1}$,

b) $X_n(s) := s^{(2^n)}$, $Y_n(s) := s^{(2^{n+1})}$,

c) $X(s) := s$, $Y_n(s) := s^n$.

**2.11.18.** Assume that $X, Y : S \to \mathbb{R}$ are discrete random variables with joint density $f$ such that $f(1, 1) = f(1, 2) = f(3, 3) = 1/10$, $f(1, 3) = f(2, 2) = 2/10$, $f(2, 3) = 3/10$, and $f(x, y) = 0$ otherwise. Find $\rho(X, Y)$.

**2.11.19.** What is the value of $\rho(X, Y)$ under each of the conditions listed below?

a) $Y = -X$,

b) $2X - 3Y = 1$,

c) $X = 4 - 3Y$.

**2.11.20.** Find $Cov(X, Y)$ for each of the joint densities given below. *Hint:* consider in each case whether the solution can be found without calculation.

a) $f(x, y) = \begin{cases} 1 & \text{if } (x, y) \in [0, 1] \times [0, 1] \\ 0 & \text{otherwise} \end{cases}$

b) $f(x, y) = \begin{cases} 2 & \text{if } 0 \leq y \leq x \leq 1 \\ 0 & \text{otherwise} \end{cases}$

c) $f(x, y) = e^{-(x^2 + y^2)/2}/(2\pi)$

**2.11.21.** For which of the curves $c(t) = (X(t), Y(t))$ in Figure 2.20 is $\rho(X, Y)$ largest and for which is it smallest? Explain your answer. Note: you may assume that in each diagram it is the case that any two curve segments of equal length represent events of equal probability in the underlying sample space from which the values $t$ are drawn.

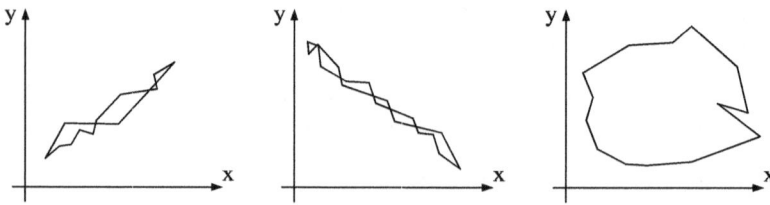

Figure 2.20: the curves $c = (X, Y)$.

**2.11.22.** Let $S := \{1, 2, 3, 4\}$ and let $P(\{k\}) := 1/4$ for all $k \in S$. Find two random variables $X, Y : S \to \mathbb{R}$ such that $\sigma(X), \sigma(Y) > 0$, $|X(k)| = |Y(k)|$ for all $k \in S$, and $\rho(X, Y) = 0$.

**2.11.23.** Let $X(s) := \sin(s)$ and $Y_n(s) := \sin(ns)$ for all $n \in \mathbb{N}$ and all $s \in [0, 2\pi]$. Show that $Cov(X, Y_n) = 0$ for all $n > 1$, where the density on $S = [0, 2\pi]$ is assumed to be uniform as usual. Are $X$ and $Y_n$ independent for any $n > 1$?

**2.11.24.** For the random variables

$$X_\alpha(s) := 3\cos(\alpha)\cos(s) - \sin(\alpha)\sin(s)$$

and

$$Y_\alpha(s) := 3\sin(\alpha)\cos(s) + \cos(\alpha)\sin(s)$$

defined on $S = [0, 2\pi]$ do the following:

**a)** Explain what the curve in $\mathbb{R}^2$ looks like—in dependence on $\alpha$—that is described by the parameterization $c_\alpha(s) = (X_\alpha(s), Y_\alpha(s))$.

**b)** Assuming that the density on $S = [0, 2\pi]$ is uniform, find $R(\alpha) := \rho(X_\alpha, Y_\alpha)$ for all $\alpha \in [0, 2\pi]$.

**c)** Plot the graph of $R$ on $[0, 2\pi]$.

**d)** Determine the minimal and maximal values of $R$.

# Chapter 3

# Limit Laws of Probability

## 3.1 The Weak Law of Large Numbers (Binomial Case)

According to Theorem 2.9.11, the variance of the average of independent random variables with common variance $\sigma^2$ converges to zero as the number of random variables that form the average increases to infinity. In particular, if the random variables are defined as in Example 2.9.9, then the sum of these random variables is a binomial random variable with parameters $n$ and $p$, and the attendant conclusion therefore is that the variance of a binomial random variable divided by $n$ converges to zero as $n$ tends to infinity. Consequently, since the variance is a measure for the deviation from the mean, it is natural to expect that the probability for a binomial random variable divided by $n$ to deviate from its mean $np/n = p$ by more than a given fixed margin $\varepsilon > 0$ converges to zero as $n$ tends to infinity. In other words, for all $\varepsilon > 0$ it ought to be the case that the probability

$$P(|X/n - p| > \varepsilon) = \sum_{\substack{k=0 \\ |k/n-p|>\varepsilon}}^{n} \binom{n}{k} p^k (1-p)^{n-k} \tag{3.1}$$

converges to zero as $n$ tends to infinity, or equivalently, that the complementary probability

$$1 - P(|X/n - p| > \varepsilon) = P(|X/n - p| \leq \varepsilon) = \sum_{\substack{k=0 \\ |k/n-p|\leq\varepsilon}}^{n} \binom{n}{k} p^k (1-p)^{n-k}$$

converges to one. Furthermore, since Theorem 2.9.11 was formulated for arbitrary independent random variables, an analogous conclusion ought to be valid as well for random variables other than binomial ones. However, before we establish this latter, more general result, we will prove in the present section that the special-case binomial probability (3.1) converges to zero by directly analyzing and estimating the binomial probability terms $\binom{n}{k}p^k(1-p)^{n-k}$. Strictly speaking, of course, this prior discussion is wholly redundant. But the contrast that it forms with the section that follows is highly instructive. For it reveals how the concept of independence allows us not only to achieve far greater generality but also to substantially reduce computational complexity. That is to say, the juxtaposition of the cumbersome special-case methods that we will use below, in the proof of Theorem 3.1.1, and the far more elegant general-case methods employed in Section 3.2 nicely brings to light the power of abstraction that inheres in the notion of independence as the central, defining notion of the entire theory of probability.

**3.1.1 Theorem.** *For any $p \in [0, 1]$ and all $\varepsilon > 0$ we have*

$$\lim_{n \to \infty} \sum_{\substack{k=0 \\ |k/n-p|>\varepsilon}}^{n} \binom{n}{k}p^k(1-p)^{n-k} = 0$$

*and*

$$\lim_{n \to \infty} \sum_{\substack{k=0 \\ |k/n-p|\leq\varepsilon}}^{n} \binom{n}{k}p^k(1-p)^{n-k} = 1.$$

*Proof.* Since the two sums in the limits above add up to 1 for every $n \in \mathbb{N}$, it is sufficient to prove only the first equation. To do so we observe to begin with that

$$n! = e^{\ln(n!)} = e^{\sum_{k=1}^{n} \ln(k)} = e^{\sum_{k=2}^{n} \ln(k)} \tag{3.2}$$

for all integers $n \geq 2$. In viewing the exponent $\sum_{k=2}^{n} \ln(k)$ to be a right-sum approximation for $\int_1^n \ln(x)\,dx$, as shown Figure 3.1, we may conclude that

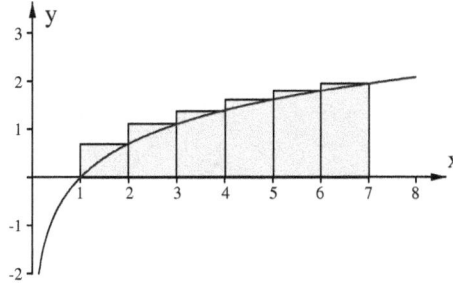

Figure 3.1: lower estimate for $\sum_{k=2}^{n} \ln(k)$ with $n = 7$.

$$\int_1^n \ln(x)\, dx \le \sum_{k=2}^n \ln(k). \qquad (3.3)$$

In other words, the integral $\int_1^n \ln(x)\, dx$ gives us a *lower* estimate for the sum $\sum_{k=2}^n \ln(k)$. Similarly, in considering $\sum_{k=2}^n \ln(k)$ to be a left-sum approximation for $\int_2^{n+1} \ln(x)\, dx$, as shown Figure 3.2, we find the *upper* estimate

$$\int_2^{n+1} \ln(x)\, dx \ge \sum_{k=2}^n \ln(k). \qquad (3.4)$$

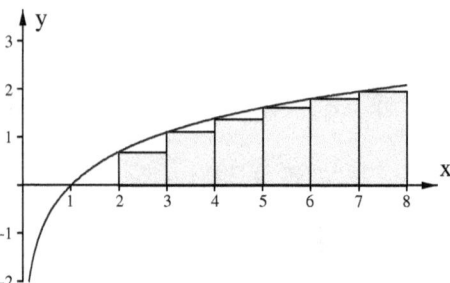

Figure 3.2: upper estimate for $\sum_{k=1}^n \ln(k)$ with $n = 7$.

Evaluating the integrals in (3.3) and (3.4) yields the combined estimate

$$n\ln(n) - n + 1 \le \sum_{k=2}^n \ln(k) \le (n+1)\ln(n+1) - (n+1) - 2\ln(2) + 2 \quad (3.5)$$

which in turn implies, in conjunction with (3.2), that

$$\frac{n^n}{e^n} < e^{n\ln(n)-n+1} \le n! \le e^{(n+1)\ln(n+1)-(n+1)-2\ln(2)+2} < \frac{(n+1)^{n+1}}{e^n}$$

for all $n \ge 2$ and trivially also for $n = 1$. Setting

$$a_n := (n+1)\left(1 + \frac{1}{n}\right)^n$$

and

$$f(x) := \frac{1}{x^x(1-x)^{1-x}},$$

it readily follows that

$$\frac{n^n}{e^n} < n! < a_n \frac{n^n}{e^n}$$

and that, by implication,

$$\binom{n}{k} = \frac{n!}{k!(n-k)!} \leq \frac{a_n n^n/e^n}{(k^k/e^k)((n-k)^{n-k}/e^{n-k})} = \frac{a_n n^n}{k^k(n-k)^{n-k}}$$

$$= \frac{a_n}{(k/n)^k(1-k/n)^{n-k}} = \frac{a_n}{((k/n)^{k/n}(1-k/n)^{1-k/n})^n} = a_n f\left(\frac{k}{n}\right)^n$$

for all $n, k \in \mathbb{N}$ with $n \geq k$. Hence

$$\sum_{\substack{k=0 \\ |k/n-p|>\varepsilon}}^{n} \binom{n}{k} p^k (1-p)^{n-k} \leq a_n \sum_{\substack{k=0 \\ |k/n-p|>\varepsilon}}^{n} f\left(\frac{k}{n}\right)^n p^k (1-p)^{n-k}.$$

Introducing the function

$$g_p(x) := \begin{cases} f(x)p^x(1-p)^{1-x} & \text{if } x \in [0,1], \\ 0 & \text{if } x \in \mathbb{R} \setminus [0,1], \end{cases}$$

we may write the inequality above in the form

$$\sum_{\substack{k=0 \\ |k/n-p|>\varepsilon}}^{n} \binom{n}{k} p^k (1-p)^{n-k} \leq a_n \sum_{\substack{k=0 \\ |k/n-p|>\varepsilon}}^{n} g_p\left(\frac{k}{n}\right)^n.$$

Furthermore, in setting the first derivative of $g_p$ equal to zero, it is easy to show (see Exercise 3.1.2 below) that $g_p$ assumes its global maximum at $x = p$ and that $g_p$ is strictly increasing on $[0, p]$ and strictly decreasing on $[p, 1]$. Consequently, in denoting by $d$ the larger of the two values $g_p(p - \varepsilon)$ and

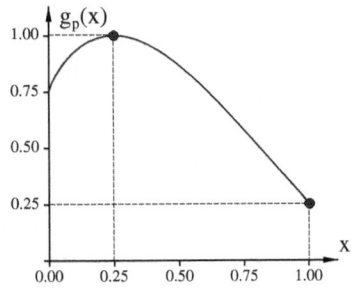

Figure 3.3: graph of $g_p(x)$ for $p = 1/4$.

$g_p(p + \varepsilon)$ (i.e., $d := \max\{g_p(p - \varepsilon), g_p(p + \varepsilon)\}$) and observing that $g_p(p) = 1$, we may infer that

$$d < 1.$$

(Note: this inequality is valid also if $p-\varepsilon \notin [0,1]$ and/or $p+\varepsilon \notin [0,1]$, because $g_p(x) = 0$ for all $x \in \mathbb{R} \smallsetminus [0,1]$.) Since $g_p$ is strictly increasing on $[0,p]$ and strictly decreasing on $[p,1]$, it follows that $g_p(x) < d$ for all $x \in [0,1]$ with $|x-p| > \varepsilon$. Hence

$$\sum_{\substack{k=0 \\ |k/n-p|>\varepsilon}}^{n} \binom{n}{k} p^k (1-p)^{n-k} \le a_n \sum_{\substack{k=0 \\ |k/n-p|>\varepsilon}}^{n} d^n \le a_n(n+1)d^n.$$

Finally, the fact that $d$ is strictly less than 1 implies that $\lim_{n\to\infty} a_n(n+1)d^n = e \lim_{n\to\infty}(n+1)^2 d^n = 0$, and that, by implication,

$$\lim_{n\to\infty} \sum_{\substack{k=0 \\ |k/n-p|>\varepsilon}}^{n} \binom{n}{k} p^k (1-p)^{n-k} = 0,$$

as desired.                                                                                $\square$

## Exercises

**3.1.2.** Show that $g_p$ assumes its global maximum at $x = p$ and that $g_p$ is strictly increasing on $[0,p]$ and strictly decreasing on $[p,1]$.

**3.1.3.** Assume that $X_n$ is a binomial random variable with parameters $n$ and $p = 1/4$. Use a computer or a calculator to find $P(|X_n/n - p| \le 0.1)$ for $n = 40$, 400, and 4000. What do you notice?

## 3.2    The Weak Law of Large Numbers (General Case)

Since the variance is a measure for the deviation from the mean, it is natural to expect that the probability $P(|X - \mu| > \varepsilon)$, to which we alluded in the preceding section, increases as $\sigma^2$ increases. After all, if $\sigma^2 = E((X - \mu)^2)$ is large, then $|X - \mu| = \sqrt{(X - \mu)^2}$ is large on average, and the likelihood for it to be greater than a given value $\varepsilon > 0$ should therefore be larger than in the alternative case where $|X - \mu|$ on average is small. Consequently, since the probability $P(|X - \mu| > \varepsilon)$ increases as $\varepsilon$ decreases (this is trivial), it is not altogether surprising to learn that the inequality

$$P(|X - \mu| > \varepsilon) \le \frac{\sigma^2}{\varepsilon^2}$$

is valid for all random variables $X : S \to \mathbb{R}$ because the term $\sigma^2/\varepsilon^2$ increases as $\sigma^2$ increases and as $\varepsilon$ decreases. Thus we have the following theorem:

**3.2.1 Theorem.** *(Chebyshev's Inequality) If $X : S \to \mathbb{R}$ is a random variable with mean $\mu$ and variance $\sigma^2$, then*

$$P(|X - \mu| > \varepsilon) \leq \frac{\sigma^2}{\varepsilon^2}$$

*and*

$$P(|X - \mu| \leq \varepsilon) \geq 1 - \frac{\sigma^2}{\varepsilon^2}$$

*for all $\varepsilon > 0$.*

*Proof.* Let $\varepsilon$ be an arbitrary positive number. If $X$ is discrete, then

$$\varepsilon^2 P(|X - \mu| > \varepsilon) = \sum_{\substack{x \in R(X) \\ |x-\mu|>\varepsilon}} \varepsilon^2 f(x) = \sum_{\substack{x \in R(X) \\ x<\mu-\varepsilon}} \varepsilon^2 f(x) + \sum_{\substack{x \in R(X) \\ x>\mu+\varepsilon}} \varepsilon^2 f(x)$$

$$\leq \sum_{\substack{x \in R(X) \\ x<\mu-\varepsilon}} (x-\mu)^2 f(x) + \sum_{\substack{x \in R(X) \\ x>\mu+\varepsilon}} (x-\mu)^2 f(x)$$

$$\leq \sum_{x \in R(X)} (x-\mu)^2 f(x) = E((X-\mu)^2) = \mathrm{Var}(X) = \sigma^2$$

and, similarly, if $X$ is continuous then

$$\varepsilon^2 P(|X - \mu| > \varepsilon) = \int_{-\infty}^{\mu-\varepsilon} \varepsilon^2 f(x)\,dx + \int_{\mu+\varepsilon}^{\infty} \varepsilon^2 f(x)\,dx$$

$$\leq \int_{-\infty}^{\mu-\varepsilon} (x-\mu)^2 f(x)\,dx + \int_{\mu+\varepsilon}^{\infty} (x-\mu)^2 f(x)\,dx$$

$$\leq \int_{-\infty}^{\infty} (x-\mu)^2 f(x)\,dx = E((X-\mu)^2) = \mathrm{Var}(X) = \sigma^2.$$

Hence

$$P(|X - \mu| > \varepsilon) \leq \frac{\sigma^2}{\varepsilon^2},$$

and, by implication,

$$P(|X - \mu| \leq \varepsilon) = 1 - P(|X - \mu| > \varepsilon) \geq 1 - \frac{\sigma^2}{\varepsilon^2},$$

as desired. $\qquad\square$

Chebyshev's inequality, as established above, applies equally to all random variables and is therefore universally applicable. However, as it turns out, the price we have to pay for this very broad applicability is lack of precision. That is to say, it often is the case that estimates better than the Chebyshev estimate can be found when additional information about the distribution of $X$ is somehow available. An example of this kind is provided by the proposition that follows.

**3.2.2 Proposition.** *If $X : S \to \mathbb{R}$ is a normal random variable (the emphasis is on 'normal') with mean $\mu$ and variance $\sigma^2$, then*

$$P(|X - \mu| > \varepsilon) \leq 1 - \sqrt{1 - e^{-\varepsilon^2/(2\sigma^2)}}$$

*and*

$$P(|X - \mu| \leq \varepsilon) \geq \sqrt{1 - e^{-\varepsilon^2/(2\sigma^2)}}$$

*for all $\varepsilon > 0$.*

*Proof.* Since the latter inequality evidently implies the former, it is sufficient to prove only the latter. So let $\varepsilon$ be an arbitrary value greater than zero and let $D \subset \mathbb{R}^2$ be the disc of radius $\varepsilon/\sigma$ that is centered at the origin, i.e.,

$$D = \{(x, y) \in \mathbb{R}^2 \mid x^2 + y^2 < \varepsilon^2/\sigma^2\}.$$

Then

$$P(|X - \mu| \leq \varepsilon) = P(\mu - \varepsilon \leq X \leq \mu + \varepsilon) = \int_{\mu-\varepsilon}^{\mu+\varepsilon} \frac{e^{-(x-\mu)^2/(2\sigma^2)}}{\sqrt{2\pi}\sigma}\, dx$$

$$= \int_{-\varepsilon/\sigma}^{\varepsilon/\sigma} \frac{e^{-x^2/2}}{\sqrt{2\pi}}\, dx = \sqrt{\int_{-\varepsilon/\sigma}^{\varepsilon/\sigma} \frac{e^{-x^2/2}}{\sqrt{2\pi}}\, dx \int_{-\varepsilon/\sigma}^{\varepsilon/\sigma} \frac{e^{-y^2/2}}{\sqrt{2\pi}}\, dy}$$

$$= \sqrt{\int_{-\varepsilon/\sigma}^{\varepsilon/\sigma} \int_{-\varepsilon/\sigma}^{\varepsilon/\sigma} \frac{e^{-(x^2+y^2)/2}}{2\pi}\, dx\, dy} \geq \sqrt{\int_{D} \frac{e^{-(x^2+y^2)/2}}{2\pi}\, dx\, dy}$$

$$= \sqrt{\int_{0}^{2\pi} \int_{0}^{\varepsilon/\sigma} \frac{e^{-r^2/2}}{2\pi} r\, dr\, d\theta} = \sqrt{\int_{0}^{\varepsilon/\sigma} r e^{-r^2/2}\, dr}$$

$$= \sqrt{1 - e^{-\varepsilon^2/(2\sigma^2)}},$$

as desired.                                                                      □

**3.2.3 Example.** Assume that $X : S \to \mathbb{R}$ is a random variable with $\text{Var}(X) = 1/16$. Then Chebyshev's inequality implies that

$$P(|X - \mu| \leq 1/2) \geq 1 - \frac{1/16}{(1/2)^2} = \frac{3}{4} = 0.75.$$

By contrast, if we assume in addition that $X$ is normal, then Proposition 3.2.2 allows us to infer that

$$P(|X - \mu| \leq 1/2) \geq \sqrt{1 - e^{-(1/2)^2/(2/16)}} = \sqrt{1 - e^{-2}} \approx 0.93,$$

which is a considerably better lower estimate. As a matter of course, in this latter case, where $X$ is known to be normal, we can also directly calculate the value of $P(|X - \mu| < 1/2)$ by means of a standard normal integral (and the 68-95-99.7 rule):

$$P(|X - \mu| \leq 1/2) = \int_{\mu-1/2}^{\mu+1/2} \frac{e^{-(x-\mu)^2/(2/16)}}{\sqrt{2\pi}/4} \, dx = \int_{-(1/2)/(1/4)}^{(1/2)/(1/4)} \frac{e^{-x^2/2}}{\sqrt{2\pi}} \, dx$$

$$= \int_{-2}^{2} \frac{e^{-x^2/2}}{\sqrt{2\pi}} \, dx \approx 0.95.$$

This possibility of directly calculating $P(|X - \mu| < \varepsilon)$ in the case where $X$ is normal somewhat diminishes the importance of Proposition 3.2.2, and we therefore chose to label it a 'proposition' rather than a 'theorem'. But an independent and valid use of this proposition we will soon encounter in the context of the weak law of large numbers in Example 3.2.7.

*Remark.* The estimates in Proposition 3.2.2 are *always* better than Chebyshev's estimates because, according to the statement of Exercise 3.2.8 below it is the case that

$$\sqrt{1 - e^{-x/2}} > 1 - 1/x$$

for all $x > 0$. So as we replace $x$ in this inequality by $\varepsilon^2/\sigma^2$, we find that universally the lower bound in Chebyshev's inequality is smaller than the lower bound given in Proposition 3.2.2.

**3.2.4 Theorem.** *(Weak Law of Large Numbers) If $(S_n)_{n\in\mathbb{N}}$ is a sequence of sample spaces and if for every $n \in \mathbb{N}$ we are given independent random variables*

$$X_{n,1}, \ldots, X_{n,n} : S_n \to \mathbb{R}$$

*with common variance $\sigma^2$ (that is, $\mathrm{Var}(X_{n,k}) = \sigma^2$ for all $n \in \mathbb{N}$ and all $k \in \{1, \ldots, n\}$), then*

$$P\left(\left|\frac{1}{n}\sum_{k=1}^{n} X_{n,k} - \mu\right| > \varepsilon\right) \leq \frac{\sigma^2}{\varepsilon^2 n},$$

$$P\left(\left|\frac{1}{n}\sum_{k=1}^{n} X_{n,k} - \mu\right| \leq \varepsilon\right) \geq 1 - \frac{\sigma^2}{\varepsilon^2 n}$$

*for all $\varepsilon > 0$ and all $n \in \mathbb{N}$, and, by implication,*

$$\lim_{n\to\infty} P\left(\left|\frac{1}{n}\sum_{k=1}^{n} X_{n,k} - \mu\right| > \varepsilon\right) = 0$$

*and*

$$\lim_{n\to\infty} P\left(\left|\frac{1}{n}\sum_{k=1}^{n} X_{n,k} - \mu\right| \leq \varepsilon\right) = 1$$

*for all $\varepsilon > 0$.*

*Proof.* Setting

$$X_n := \frac{1}{n} \sum_{k=1}^{n} X_{n,k},$$

it follows that

$$E(X_n) = \frac{1}{n} E\left(\sum_{k=1}^{n} X_{n,k}\right) = \frac{1}{n} \sum_{k=1}^{n} \mu = \mu$$

and, according to Theorems 2.9.8 and 2.3.7b, it is the case that

$$\mathrm{Var}(X_n) = \frac{1}{n^2} \mathrm{Var}\left(\sum_{k=1}^{n} X_{n,k}\right) = \frac{1}{n^2} \sum_{k=1}^{n} \sigma^2 = \frac{\sigma^2}{n}$$

Consequently, Chebyshev's inequality—applied to $X_n$ in place of $X$—allows us to infer that

$$P\left(\left|\frac{1}{n} \sum_{k=1}^{n} X_{n,k} - \mu\right| > \varepsilon\right) = P\left(|X_n - \mu| > \varepsilon\right) \leq \frac{\sigma^2}{\varepsilon^2 n},$$

and that, by implication,

$$\lim_{n \to \infty} P\left(\left|\frac{1}{n} \sum_{k=1}^{n} X_{n,k} - \mu\right| > \varepsilon\right) = 0.$$

Hence it also is the case that

$$P\left(\left|\frac{1}{n} \sum_{k=1}^{n} X_{n,k} - \mu\right| \leq \varepsilon\right) \geq 1 - \frac{\sigma^2}{\varepsilon^2 n},$$

and that

$$\lim_{n \to \infty} P\left(\left|\frac{1}{n} \sum_{k=1}^{n} X_{n,k} - \mu\right| \leq \varepsilon\right) = 1$$

because

$$P\left(\left|\frac{1}{n} \sum_{k=1}^{n} X_{n,k} - \mu\right| > \varepsilon\right) + P\left(\left|\frac{1}{n} \sum_{k=1}^{n} X_{n,k} - \mu\right| \leq \varepsilon\right) = 1$$

for all $\varepsilon > 0$ and all $n \in \mathbb{N}$.                                            $\square$

In order to improve the estimates provided by in the preceding theorem by means of Proposition 3.2.2—in the case where the random variables $X_{n,k}$ are not only independent but also normal—we need to establish the following lemma (see also Exercise 2.9.19):

**3.2.5 Lemma.** *If $X$ is normal with mean $\mu$ and variance $\sigma^2$, then, for all $\alpha \in \mathbb{R} \setminus \{0\}$, the random variable $\alpha X$ is normal as well, and its mean is $\alpha\mu$ and its variance is $\alpha^2\sigma^2$.*

*Proof.* For any $x \in \mathbb{R}$ the value of the cumulative density of $\alpha X$ is $P(\alpha X \le x)$. Thus we may distinguish two cases:

*Case 1: $\alpha > 0$.* Then

$$P(\alpha X \le x) = P(X \le x/\alpha) = \int_{-\infty}^{x/\alpha} \frac{e^{-(u-\mu)^2/(2\sigma^2)}}{\sqrt{2\pi}\sigma} \, du,$$

and therefore, using the fundamental theorem of calculus in conjunction with the chain rule, we find that the density of $\alpha X$ is

$$\frac{d}{dx} \int_{-\infty}^{x/\alpha} \frac{e^{-(u-\mu)^2/(2\sigma^2)}}{\sqrt{2\pi}\sigma} \, du = \frac{e^{-(x/\alpha-\mu)^2/(2\sigma^2)}}{\sqrt{2\pi}\sigma} \cdot \frac{1}{\alpha} = \frac{e^{-(x-\alpha\mu)^2/(2\alpha^2\sigma^2)}}{\sqrt{2\pi}\alpha\sigma}.$$

Conequently, $\alpha X$ is normal with mean $\alpha\mu$ and variance $\alpha^2\sigma^2$, as desired.

*Case 2: $\alpha < 0$.* Then

$$P(\alpha X \le x) = P(X \ge x/\alpha) = \int_{x/\alpha}^{\infty} \frac{e^{-(u-\mu)^2/(2\sigma^2)}}{\sqrt{2\pi}\sigma} \, du$$

$$= -\int_{\infty}^{x/\alpha} \frac{e^{-(u-\mu)^2/(2\sigma^2)}}{\sqrt{2\pi}\sigma} \, du,$$

and, by implication, the density of $\alpha X$ is

$$-\frac{d}{dx} \int_{\infty}^{x/\alpha} \frac{e^{-(u-\mu)^2/(2\sigma^2)}}{\sqrt{2\pi}\sigma} \, du = -\frac{e^{-(x/\alpha-\mu)^2/(2\sigma^2)}}{\sqrt{2\pi}\sigma} \cdot \frac{1}{\alpha} = \frac{e^{-(x-\alpha\mu)^2/(2\alpha^2\sigma^2)}}{\sqrt{2\pi}|\alpha|\sigma}.$$

Thus we find as above that $\alpha X$ is normal with mean $\alpha\mu$ and variance $\alpha^2\sigma^2$. $\square$

**3.2.6 Proposition.** *If $X_1, \ldots, X_n$ are independent normal random variables with common mean $\mu$ and common variance $\sigma^2$, then*

$$\frac{1}{n} \sum_{k=1}^{n} X_k$$

*is normal with mean $\mu$ and variance $\sigma^2/n$ and*

$$P\left(\left|\frac{1}{n} \sum_{k=1}^{n} X_k - \mu\right| > \varepsilon\right) \le 1 - \sqrt{1 - e^{-n\varepsilon^2/(2\sigma^2)}}$$

*and*

$$P\left(\left|\frac{1}{n} \sum_{k=1}^{n} X_k - \mu\right| \le \varepsilon\right) \ge \sqrt{1 - e^{-n\varepsilon^2/(2\sigma^2)}}$$

*for all $\varepsilon > 0$.*

*Proof.* According to Theorem 2.9.12, the sum $\sum_{k=1}^{n} X_k$ is normal with mean $n\mu$ and variance $n\sigma^2$, and therefore, Lemma 3.2.5 (with $\alpha = 1/n$) implies that the random variable

$$X := \frac{1}{n} \sum_{k=1}^{n} X_k$$

is normal with mean $\mu$ and variance $\sigma^2/n$. Given this observation, Theorem 3.2.6 directly follows from Theorem 3.2.2 with $\sigma^2/n$ in place of $\sigma^2$. □

**3.2.7 Example.** Assume that for every $k \in \mathbb{N}$ we are given a random variable $X_k : S \to \mathbb{R}$ with mean $\mu$ and variance $\sigma^2 = 1/4$ such that for all $n \in \mathbb{N}$ the random variables $X_1, \ldots, X_n$ are independent. Given this information, we wish to determine the least value for $n$ for which it is the case that

$$P\left( \left| \frac{1}{n} \sum_{k=1}^{n} X_k - \mu \right| \leq 0.1 \right) \geq 0.95. \tag{3.6}$$

Using the second inequality in Theorem 3.2.4, it follows that (3.6) is satisfied if

$$1 - \frac{1/4}{0.1^2 n} = 1 - \frac{25}{n} \geq 0.95$$

or, equivalently, if $n \geq 25/0.05 = 500$. If we assume in addition that each $X_k$ is normal, then we may use Proposition 3.2.6 to conclude that (3.6) is satisfied whenever

$$\sqrt{1 - e^{-n0.1^2/(2/4)}} = \sqrt{1 - e^{-n/50}} \geq 0.95$$

or, equivalently, whenever

$$n \geq -50 \ln(1 - 0.95^2) \approx 116.4.$$

So in using the universally valid Chebyshev estimate, we find the minimal value for $n$ to be 500, and as we impose the additional assumption of normality, this minimal value is reduced to 117.

To generalize this result, we may naturally replace 0.1 by $\varepsilon$, 1/4 by $\sigma^2$, and 0.95 by some arbitrary value $q \in (0, 1)$. That is to say, the problem that we pose is to determine the least integer $n$ that satisfies the inequality

$$P\left( \left| \frac{1}{n} \sum_{k=1}^{n} X_k - \mu \right| \leq \varepsilon \right) \geq q.$$

Using Theorem 3.2.4, we find that

$$1 - \frac{\sigma^2}{n\varepsilon^2} \geq q$$

or equivalently that

$$n \geq \frac{\sigma^2}{\varepsilon^2(1-q)}, \tag{3.7}$$

and using Proposition 3.2.6, in the more restrictive normal case, it follows that

$$\sqrt{1 - e^{-n\varepsilon^2/(2\sigma)^2}} \geq q$$

or equivalently that

$$n \geq -\frac{2\sigma^2 \ln(1-q^2)}{\varepsilon^2}. \tag{3.8}$$

(Note: $\ln(1 - q^2)$ is negative for all $q \in (0, 1)$, and therefore, $-\ln(1 - q^2)$ is positive.) In order to understand why the lower bound in (3.8) is indeed substantially smaller than the one in (3.7), it is instructive to examine the quotient of these two lower bounds as $q$ approaches 1 from below. To do so, we use L'Hôpital's rule to infer that

$$\lim_{q \to 1^-} -\frac{2\sigma^2 \ln(1-q^2)/\varepsilon^2}{\sigma^2/(\varepsilon^2(1-q))} = \lim_{q \to 1^-} -\frac{2\ln(1-q^2)}{1/(1-q)}$$

$$= \lim_{q \to 1^-} -\frac{2(\ln(1-q) + \ln(1+q))}{1/(1-q)}$$

$$= \lim_{q \to 1^-} -\frac{2\ln(1-q)}{1/(1-q)} = \lim_{x \to 0^+} -\frac{2\ln(x)}{1/x}$$

$$= \lim_{x \to 0^+} \frac{2/x}{1/x^2} = 0.$$

This shows that for values $q$ that are close to one, the lower bound in (3.8) is vanishingly small compared to the lower bound in (3.7). Finally, concerning the question of the independent usefulness of Proposition 3.2.2 (and Proposition 3.2.6 thereby), to which we alluded earlier in Example 3.2.3, we wish to point out that a direct evaluation of the probability

$$P\left( \left| \frac{1}{n} \sum_{k=1}^{n} X_k - \mu \right| \leq \varepsilon \right)$$

by means of a normal integral does not yield an estimate analogous to (3.8) without some additional arguments of the type that we employed in the proof of Proposition 3.2.2. For since there is no closed-form antiderivative of $e^{-x^2}$ (there is a series-form antiderivative but the series cannot be reduced to a simple formula), it is impossible to directly solve for $n$ the inequality

$$P\left( \left| \frac{1}{n} \sum_{k=1}^{n} X_k - \mu \right| \leq \varepsilon \right) = \int_{\mu-\varepsilon}^{\mu+\varepsilon} \frac{e^{-n(x-\mu)^2/(2\sigma^2)}}{\sqrt{2\pi}\sigma/\sqrt{n}} \, dx$$

$$= \int_{-\varepsilon\sqrt{n}/\sigma}^{\varepsilon\sqrt{n}/\sigma} \frac{e^{-x^2/2}}{\sqrt{2\pi}} \, dx \geq q.$$

Consequently, Proposition 3.2.2 is not rendered useless by the fact that the density of the average of the random variables $X_1, \ldots, X_n$ is known to be normal with mean $\mu$ and variance $\sigma^2/n$.

# Exercises

**3.2.8.** Show that $\sqrt{1 - e^{-x/2}} > 1 - 1/x$ for all $x > 0$. *Hint:* use a tangent line approximation to show that $e^x \geq 1 + x$ for all $x \in \mathbb{R}$ and then use this fact to prove the given inequality for all $x \geq 1$.

**3.2.9.** Let $\varepsilon$ be a given positive number and assume that $X_1, \ldots, X_n$ are independent random variables with common mean $\mu$. Use Chebyshev's inequality to show that

$$P\left(\left|\frac{1}{n}\sum_{k=1}^{n} X_k - \mu\right| \leq \varepsilon\right) \geq 1 - \frac{1}{\varepsilon^2 n^2}\sum_{k=1}^{n} \sigma_k^2,$$

where $\sigma_k^2 = Var(X_k)$ for all $k \in \{1, \ldots, n\}$.

**3.2.10.** Let $\varepsilon$ be a given positive number and assume that $(X_k)_{k \in \mathbb{N}}$ is a sequence of random variables with common mean $\mu$ such that $\sigma_k \leq \sqrt{k}^\alpha$ for some $\alpha \in (0, 1)$ and all $k \in \mathbb{N}$ and such that $X_1, \ldots, X_k$ are independent for all $k \in \mathbb{N}$. Use the result of Exercise 3.2.9 to show that

$$\lim_{n \to \infty} P\left(\left|\frac{1}{n}\sum_{k=1}^{n} X_k - \mu\right| \leq \varepsilon\right) = 1.$$

**3.2.11.** Explain why the conclusion in Exercise 3.2.10 remains valid if we assume that $\sigma_k \leq \sqrt{k/\ln(k)}$ for all $k \geq 2$.

**3.2.12.** Assume that $X$ is a normal random variable with mean $\mu$ and standard deviation $\sigma = 1/20$. Use Chebyshev's inequality to find a lower estimate for $P(|X - \mu| \leq 0.1)$.

**3.2.13.** Use Proposition 3.2.2 to find a lower estimate for the same probability as in Exercise 3.2.12.

**3.2.14.** Assume that $(X_k)_{k \in \mathbb{N}}$ is a sequence random variables with common mean $\mu = 2$ and common variance $\sigma^2 \leq 1$ such that $X_1, \ldots, X_k$ are independent for all $k \in \mathbb{N}$. Would you be surprised to learn—in light of these assumptions and in light of Chebyshev's inequality—that the sample average of the first 1,000,000 trials came out to be 2.003? What if the same sample average had been found after 1,000,000,000 trials?

**3.2.15.** Assume that $(X_k)_{k \in \mathbb{N}}$ is a sequence of *normal* random variables with common mean $\mu = 2$ and common variance $\sigma^2 \leq 1$ such that $X_1, \ldots, X_k$ are independent for all $k \in \mathbb{N}$. Would you be surprised to learn—in light of these assumptions and in light of Proposition 3.2.6—that the sample average of the first 1,000,000 trials came out to be 2.003?

# 3.3　The Strong Law of Large Numbers

Given that there is a *weak* law of large numbers, the question naturally arises as to whether there also is a *strong* law—and the answer is "yes." Unfortunately, though, the conceptual tools that are at our disposal in this introductory text are a bit too limited to explain the matter fully. But roughly speaking, what is at issue here is the distinction between convergence in probability, as predicted by the weak law, and convergence at individual points $s \in S$. The weak law in essence says that, given a sequence of independent random variables $X_k$ with common mean $\mu$ and common variance $\sigma^2$, the *probability* for the random variable

$$\frac{1}{n} \sum_{k=1}^{n} X_k$$

to differ from $\mu$ by more than a fixed margin $\varepsilon > 0$ converges to zero as $n$ tends to infinity, but it doesn't say that the difference between $\mu$ and the average

$$\frac{1}{n} \sum_{k=1}^{n} X_k(s) \tag{3.9}$$

at any given point $s$ in the sample space $S$ converges to zero as well. As it turns out, weak convergence in probability does not in general imply pointwise convergence at all or 'almost all' points $s \in S$ (see Example 3.3.8 below), but given the present assumptions of independence and uniformity in mean and variance, it actually is possible to establish the following theorem (see [A], p.274, for the proof):

**3.3.1 Theorem.** *(Strong Law of Large Numbers, Common Variance Case) Assume that for every $k \in \mathbb{N}$ we are given a random variable $X_k : S \to \mathbb{R}$ such that $X_1, \ldots, X_k$ are independent for all $k \in \mathbb{N}$ and such that $E(X_k) = \mu$ and $\mathrm{Var}(X_k) = \sigma^2$ for all $k \in \mathbb{N}$ and some $\mu, \sigma \in \mathbb{R}$ with $\sigma \geq 0$. Then there exists a set $M \subset S$ such that $P(M) = 1$ and*

$$\lim_{n \to \infty} \frac{1}{n} \sum_{k=1}^{n} X_k(s) = \mu$$

*for all $s \in M$. (Note: as we already pointed out in the statement of Theorem 2.11.10, the condition $P(M) = 1$ does not in general imply that $M = S$.)*

**3.3.2 Example.** To illustrate the significance of this theorem, it is helpful to reconsider the sample space $S$ of all infinite 01-sequences in Example 1.5.5 (i.e., $S = \{(s_k)_{k \in \mathbb{N}} \mid s_k \in \{0, 1\}$ for all $k \in \mathbb{N}\}$). In order to define for every $k \in \mathbb{N}$ a random variable $X_k : S \to \mathbb{R}$, we set

$$X_k(s) := s_k \in \{0, 1\}$$

for all $s = (s_k)_{k \in \mathbb{N}} \in S$. So each $X_k$ equals the outcome of the $k$-th coin toss in the infinite sequence of coin tosses that $s$ represents. Given this definition, it follows that the range of each $X_k$ is $\{0, 1\}$, and using the map $B : S \to [0, 1]$ and the sets $E_k$ that we introduced in Example 1.5.5, we may infer that

$$P(\{X_k = 0\}) = P(B^{-1}(E_k)) = P(E_k) = \frac{1}{2}, \qquad (3.10)$$

and similarly,

$$P(\{X_k = 1\}) = P(B^{-1}([0, 1] \setminus E_k)) = P([0, 1] \setminus E_k) = 1 - \frac{1}{2} = \frac{1}{2}. \quad (3.11)$$

Consequently,

$$\mu = E(X_k) = 0 \cdot P(\{X_k = 0\}) + 1 \cdot P(\{X_k = 1\}) = \frac{1}{2}$$

and

$$\sigma^2 = E(X_k^2) - E(X_k)^2 = 0^2 \cdot P(\{X_k = 0\}) + 1^2 \cdot P(\{X_k = 1\}) - \frac{1}{4} = \frac{1}{4}$$

for all $k \in \mathbb{N}$. Furthermore, as we combine the proven independence of the sets $E_k$ (see Example 1.5.5) with (3.10) and (3.11), it is not difficult to show (see Exercise 3.3.10 below) that the random variables $X_1, \ldots, X_k$ are independent for all $k \in \mathbb{N}$. Thus, Theorem 3.3.1 allows us to conclude that there is a set $M \subset S$ with $P(M) = 1$ (i.e., $M = B^{-1}(N)$ for some $N \subset [0, 1]$ with $P(N) = 1$) such that

$$\lim_{n \to \infty} \frac{1}{n} \sum_{k=1}^{n} X_k(s) = \lim_{n \to \infty} \frac{1}{n} \sum_{k=1}^{n} s_k = \mu = \frac{1}{2}$$

for all $s = (s_k)_{k \in \mathbb{N}} \in M$. In other words, we can be mathematically absolutely certain (because $P(M) = 1$) that as we keep flipping an unbiased coin *ad infinitum*, the number of heads over the total number of trials will approach ever more closely the ideal mean $\mu = 1/2$. Put differently, it really is impossible— in theory at least—to ever produce an actual sequence of perfectly unbiased coin tosses in which the numbers of heads and tails, in the eventual limit, are not perfectly balanced. (As a matter of course, in the actual physical universe nothing is ever perfect and no sequence of trials can ever be infinite. But then again, who cares? The whole charade of probability is anyway entirely a human mental construct and thus entirely unreal—except for the fact that the human mental world is the only world that we can ever know and that nothing, therefore, can ever be more real to us than a human mental construct.)

To continue our discussion of Theorem 3.3.1, we wish to point out that the assumption that the random variables $X_k$ have a common *finite* variance $\sigma^2$

can be dropped if we assume that all the $X_k$ have the same distribution. That is to say, what is at issue here is the general fact that the mean of a random variable can be finite while its variance is infinite (see also Exercise 2.3.20). For instance, for the random variable $X(s) := 1/\sqrt{s}$ on $S = (0, 1]$ we find that

$$E(X) = \int_0^1 \frac{1}{\sqrt{s}}\, ds = 2 < \infty,$$

but

$$\text{Var}(X) = E(X^2) - E(X)^2 = \int_0^1 \frac{1}{s}\, ds - 4 = \infty.$$

Consequently and in the light of this example, the meaning of Theorem 3.3.5 below is to assert the same pointwise convergence property as in Theorem 3.3.1 but in the case where the random variables $X_k$ are identically distributed and where, potentially, $\text{Var}(X_k) = \infty$ for all $k \in \mathbb{N}$.

**3.3.3 Definition.** Given an integer $n \in \mathbb{N}$, we say that (discrete or continuous) random variables $X_1, \ldots, X_n : S \to \mathbb{R}$ are *identically distributed* if for all suitable sets $B \subset \mathbb{R}$ it is the case that

$$P(X_1 \in B) = \cdots = P(X_n \in B)$$

(where as usual $\{X_k \in B\} := \{s \in S \mid X_k(s) \in B\}$).

*Remark.* For all practical purposes and in the context of the present exposition, the condition stated in the preceding definition simply means that the random variables $X_1, \ldots, X_n$ have the same density. For in the discrete case we may set $B_x := \{x\}$ for any given $x \in \mathbb{R}$ to conclude that

$$f(x) = P(X_1 \in B_x) = \cdots = P(X_n \in B_x) = P(X_1 = x) = \cdots = P(X_n = x)$$

is the common density of $X_1, \ldots, X_n$, and in the continuous case we may set $B_x := (-\infty, x]$ for all $x \in \mathbb{R}$ to infer that the random variables $X_1, \ldots, X_n$ have the same cumulative density and therefore also the same density because $f(x) = F'(x)$:

$$F(x) = P(X_1 \in B_x) = \cdots = P(X_n \in B_x) = P(X_1 \le x) = \cdots = P(X_n \le x).$$

Conversely, if $X_1, \ldots, X_n$ have the same density $f$, then, in the continuous case, we find that

$$\int_B f(x)\, dx = P(X_1 \in B) = \cdots = P(X_n \in B)$$

for all suitable $B \subset \mathbb{R}$, and in the discrete case, the same conclusion holds as well (see Exercise 3.3.11). Consequently, identical distribution and equal density are indeed equivalent conditions (in the context of the present exposition in which probability measures are not discussed in full generality).

Naturally, this observation also shows that the assumption of identical distribution trivially implies that all the means and all the (potentially infinite) variances of the random variables $X_k$ are equal. After all, the mean and the variance are defined in terms of the density, and, by implication, if two random variables have the same density, they also must have the same mean and the same variance. So whenever random variables $X_1, \ldots, X_n : S \to \mathbb{R}$ are identically distributed, it must be the case that

$$E(X_1) = \cdots = E(X_n)$$

and

$$\mathrm{Var}(X_1) = \cdots = \mathrm{Var}(X_n).$$

As a matter of course, the reverse conclusion is inadmissible. For it is easy to find two random variables $X$ and $Y$ that are not identically distributed but that nonetheless have the same mean and the same variance (see Exercise 3.3.12 below).

**3.3.4 Definition.** If for every $k \in \mathbb{N}$ we are given a random variable $X_k : S \to \mathbb{R}$, then $(X_k)_{k \in \mathbb{N}}$ is said to be a *sequence of i.i.d. random variables* if for every $n \in \mathbb{N}$ the random variables $X_1, \ldots, X_n$ are independent and identically distributed.

**3.3.5 Theorem.** *(Strong Law of Large Numbers, i.i.d. Case) If $(X_k)_{k \in \mathbb{N}}$ is a sequence of i.i.d. random variables with (common) finite mean $\mu$, then there exists a set $M \subset S$ such that $P(M) = 1$ and*

$$\lim_{n \to \infty} \frac{1}{n} \sum_{k=1}^{n} X_k(s) = \mu$$

*for all $s \in M$ (see [A], pp.275–277, for the proof).*

To conclude this section, we will briefly address the general theoretical question of the relation between convergence in probability and convergence at individual points. In order to do this properly, we need to make reference to the rigorous definition of a limit of a sequence (see for instance, [B1], Chapter 3), and readers who are not familiar with this definition or have no experience in working with it may therefore want to move on to the far more important topic of the Central Limit Theorem in the next and final section of this chapter.

**3.3.6 Definition.** A sequence $(a_n)_{n \in \mathbb{N}}$ of real numbers is said to converge to the *limit* $L \in \mathbb{R}$ (i.e., $L = \lim_{n \to \infty} a_n$) if for every $\delta > 0$ there exists an $N \in \mathbb{N}$ such that for all $n \geq N$ it is the case that

$$|a_n - L| \leq \delta.$$

**3.3.7 Definition.** A sequence of random variables $(X_n)_{n\in\mathbb{N}}$ on a sample space $S$ is said to converge to $L \in \mathbb{R}$ *in probability* if for all $\varepsilon > 0$ it is the case that

$$\lim_{n\to\infty} P(|X_n - L| > \varepsilon) = 0,$$

and $(X_n)_{n\in\mathbb{N}}$ is said to converge to $L$ *pointwise almost everywhere* if there exists a set $M \subset S$ of maximal probability (i.e., $P(M) = 1$) such that

$$\lim_{n\to\infty} X_n(s) = L$$

for all $s \in M$.

**3.3.8 Example.** In order to show that convergence in probability does not imply pointwise convergence almost everywhere, we will construct a sequence of random variables $(X_n)_{n\in\mathbb{N}}$ on the sample space $S = [0,1]$ that converges to zero in probability but does not converge pointwise anywhere in $S$. To do so, we define a collection of intervals $I_{m,k}$ and random variables $Y_{m,k} : S \to \mathbb{R}$ as follows: for a given $m \in \mathbb{N}$ and a given $k \in \{1,\dots,m\}$ we set

$$I_{m,k} := \left[\frac{k-1}{m}, \frac{k}{m}\right]$$

and

$$Y_{m,k}(s) := \begin{cases} 1 & \text{if } s \in I_{m,k}, \\ 0 & \text{if } s \in [0,1] \setminus I_{m,k}. \end{cases}$$

For instance, for $m = 3$ we have

$$I_{3,1} = [0, 1/3], \quad I_{3,2} = [1/3, 2/3], \quad I_{3,3} = [2/3, 1],$$

and the random variables $Y_{3,1}$, $Y_{3,2}$, and $Y_{3,3}$ are correspondingly shown in Figure 3.4. To proceed, we define the random variables $X_n$ to be successively

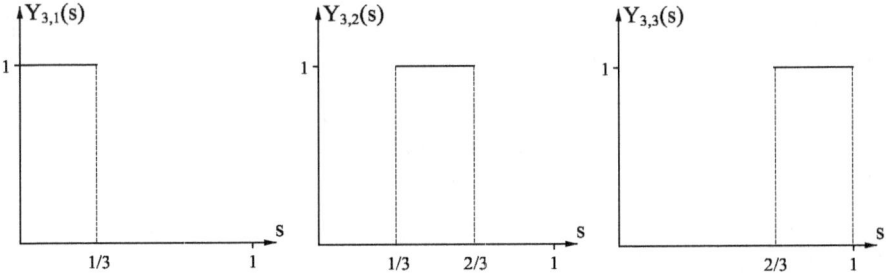

Figure 3.4: the random variables $Y_{3,1}$, $Y_{3,2}$, and $Y_{3,3}$.

equal to the random variables $Y_{m,k}$, that is,

$$X_1 := Y_{1,1},$$
$$X_2 := Y_{2,1}, \quad X_3 := Y_{2,2},$$
$$X_4 := Y_{3,1}, \quad X_5 := Y_{3,2}, \quad X_6 := X_{3,3}$$

and so forth *ad infinitum.* Since

$$P(|Y_{m,k}| > \varepsilon) = \begin{cases} 0 & \text{if } \varepsilon \geq 1, \\ P(I_{m,k}) = 1/m & \text{if } 0 < \varepsilon < 1, \end{cases}$$

and since $\lim_{m \to \infty} 1/m = 0$, the definition of the random variables $X_n$ above evidently implies that

$$\lim_{n \to \infty} P(|X_n| > \varepsilon) = 0$$

for all $\varepsilon > 0$. So the sequence $(X_n)_{n \in \mathbb{N}}$ converges to zero in probability. However, there is no point $s \in S = [0, 1]$ at which the sequence $X_n(s)$ converges to zero or any other value $L \in \mathbb{R}$ because, due to the way in which the random variables $Y_{m,k}$ are defined, the values $X_n(s)$ continue to alternate between zero and one as $n$ increases to infinity. This is so because for every $m \geq 3$ and every $s \in S$, there always exist indices $k_1, k_2 \in \{1, \ldots, m\}$ such that $s \in I_{m,k_1}$ and $s \notin I_{m,k_2}$ or, equivalently, such that $Y_{m,k_1}(s) = 1$ and $Y_{m,k_2}(s) = 0$. Consequently, if $L \in \mathbb{R}$ is a fixed given value, then either

$$|Y_{m,k_1}(s) - L| \geq 1/2$$

or

$$|Y_{m,k_2}(s) - L| \geq 1/2.$$

By implication, for $\delta := 1/2$ and for any $N \in \mathbb{N}$ we can always find an $n \geq N$ such that

$$|X_n(s) - L| \geq 1/2 = \delta.$$

In other words, according to Definition 3.3.6, there does not exist a value $L \in \mathbb{R}$ to which the sequence $(X_n(s))_{n \in \mathbb{N}}$ converges. So as we claimed, the sequence $(X_n(s))_{n \in \mathbb{N}}$ is not convergent for any $s \in S$.

**3.3.9 Theorem.** *If $(X_n)_{n \in \mathbb{N}}$ is a sequence of random variables (on a sample space $S$) that converges pointwise almost everywhere to a value $L \in \mathbb{R}$, then $(X_n)_{n \in \mathbb{N}}$ converges to $L$ in probability.*

*Proof.* Let $\varepsilon > 0$. According to Definitions 3.3.6 and 3.3.7, we need to show that for every $\delta > 0$ there exists an $N \in \mathbb{N}$ such that

$$P(|X_n - L| > \varepsilon) < \delta$$

for all $n \geq N$. To this end we pick a $\delta > 0$ and set

$$E_N := \{s \in S \mid |X_n(s) - L| \leq \varepsilon \text{ for all } n \geq N\}$$

for all $N \in \mathbb{N}$. Then, trivially,

$$E_N \subset E_{N+1} \tag{3.12}$$

for all $N \in \mathbb{N}$ because if $|X_n(s) - L| \leq \varepsilon$ for all $n \geq N$, then it also is the case that $|X_n(s) - L| \leq \varepsilon$ for all $n \geq N + 1$. Moreover, since $(X_n)_{n \in \mathbb{N}}$ is assumed to converge to $L$ pointwise almost everywhere, there exists a set $M \subset S$ such that $P(M) = 1$ and

$$\lim_{n \to \infty} X_n(s) = L$$

for all $s \in M$. Consequently, according to Definition 3.3.6, we can find for every $s \in M$ an $N \in \mathbb{N}$ such that

$$|X_n(s) - L| < \varepsilon$$

for all $n \geq N$. Thus, for every $s \in M$ there exists an $N \in \mathbb{N}$ such that $s \in E_N$, and this in turn implies that

$$M \subset \bigcup_{N=1}^{\infty} E_N.$$

Hence

$$1 = P\left(\bigcup_{N=1}^{\infty} E_N\right) = \lim_{N \to \infty} P(E_N) \quad \text{(by (3.12) and Exercise 3.3.13)},$$

and therefore, there must exist an $N \in \mathbb{N}$ such that

$$P(E_N) > 1 - \delta.$$

Furthermore, if we now pick an arbitrary $n \geq N$, then, trivially,

$$E_N \subset \{s \in S \mid |X_n(s) - L| \leq \varepsilon\}$$

(because if $|X_n(s) - L|$ is less than or equal to $\varepsilon$ for all $n \geq N$, then it is also less than or equal to $\varepsilon$ for one particular $n \geq N$), and, by implication,

$$P(E_N) \leq P(|X_n - L| \leq \varepsilon).$$

Thus

$$P(|X_n - L| > \varepsilon) = 1 - P(|X_n - L| \leq \varepsilon) \leq 1 - P(E_N) < 1 - (1 - \delta) = \delta,$$

as desired. $\qquad\square$

## Exercises

**3.3.10.** Show that the random variables $X_k$ defined in Example 3.3.2 satisfy the requirement in Theorem 3.3.1 that $X_1, \ldots, X_n$ be independent for all $n \in \mathbb{N}$.

**3.3.11.** Assume that $X_1, \ldots, X_n$ are discrete random variables with common density $f$. Show that $P(X_1 \in B) = \cdots = P(X_n \in B)$ for all (suitable) sets $B \subset \mathbb{R}$.

**3.3.12.** Find two discrete random variables $X, Y : [0,1] \to \mathbb{R}$ that are not identically distributed but have the same mean and variance. As usual and for simplicity you may assume that the probability distribution on the sample space $S = [0,1]$ is uniform.

**3.3.13.** Assume that $(E_n)_{n \in \mathbb{N}}$ is a sequence of events in a sample space $S$ such that $E_n \subset E_{n+1}$ for all $k \in \mathbb{N}$. Show that

$$P \left( \bigcup_{n=1}^{\infty} E_n \right) = \lim_{n \to \infty} P(E_n).$$

*Hint.* Define $F_1 := E_1$ and $F_n := E_n \setminus E_{n-1}$ for all $n \in \mathbb{N} \setminus \{1\}$ and show that $(F_n)_{n \in \mathbb{N}}$ is a sequence of mutually exclusive events and that

$$\bigcup_{n=1}^{\infty} E_n = \bigcup_{n=1}^{\infty} F_n.$$

Then apply the third axiom of probability to $(F_n)_{n \in \mathbb{N}}$.

**3.3.14.** Let $S$ be a finite sample space such that $P(s) > 0$ for all $s \in S$ and let $(X_n)_{n \in \mathbb{N}}$ be a sequence of random variables that converges to zero in probability. Show that $(X_n(s))_{n \in \mathbb{N}}$ converges to zero for all $s \in S$.

**3.3.15.** Let $P(E) := \int_E z(s) \, ds$ for all (suitable) events $E \subset S = \mathbb{R}$ (where $z$ is the standard normal density), and let $X_n(s) := s^2/n$ for all $n \in \mathbb{N}$ and all $s \in \mathbb{R}$. Use Theorem 3.3.9, to show that $(X_n)_{n \in \mathbb{N}}$ converges to zero in probability.

**3.3.16.** Assume that the probability density on the sample space $S = [0,1]$ is the uniform density and for any $n \in \mathbb{N}$ let

$$X_n(s) := \begin{cases} 1 & \text{if } s \in \{1/n, 2/n, \ldots, n/n\}, \\ 0 & \text{if } s \in [0,1] \setminus \{1/n, 2/n, \ldots, n/n\}. \end{cases}$$

Explain why the sequence $(X_n)_{n \in \mathbb{N}}$ converges to zero in probability.

# 3.4 The Central Limit Theorem

Having introduced the notion of identical distribution in the preceding section, we are now ready to state and discuss in some more detail the so-called *Central Limit Theorem* to which we earlier referred in the introduction to Section 2.7. Given the truly central place that this theorem occupies in the theory of probability, it seems proper to remind ourselves exactly what this theorem's assertion really is: whenever we perform a sequence of random trials independently, using always the same physical or algorithmic procedure in the process, the averages of the output values of the random variables that represent these trials will be distributed approximately normally, and in the limit, as the number of terms in these averages tends to infinity, the error in the approximation decreases to zero. So whenever our object of study is a sequence of random values that are independently generated by the same kind of random-trial procedure—so as to guarantee thereby that the corresponding random variables are identically distributed—the Central Limit Theorem can be applied. In other words, the range of applications that the Central Limit Theorem covers is truly enormous, and it is for this very reason that the normal distribution is rightly considered to be the most important distribution in the entire field of probability and statistics.

**3.4.1 Theorem.** *(The Central Limit Theorem) If $(X_k)_{k\in\mathbb{N}}$ is a sequence of i.i.d. random variables, then for all sufficiently large $n$ the random variable*

$$Y_n := \sum_{k=1}^{n} X_k$$

*is approximately normal with mean $n\mu$ and variance $n\sigma^2$ (in the sense defined below), where $\mu$ and $\sigma^2$ are the common mean and the common variance, respectively, of the i.i.d. random variables $X_k$ (see the remark on p.159f). By implication, the random variable $\overline{X}_n := Y_n/n$ is approximately normal with mean $\mu$ and variance $\sigma^2/n$, and the random variable*

$$Z_n := \frac{\overline{X}_n - \mu}{\sigma/\sqrt{n}}$$

*is approximately standard normal (by Corollary 2.6.7). More precisely, this means that for all $a, b \in \mathbb{R}$ with $a < b$ it is the case that*

$$\lim_{n\to\infty} P\left(\frac{\overline{X}_n - \mu}{\sigma/\sqrt{n}} \in [a, b]\right) = \int_a^b \frac{e^{-x^2/2}}{\sqrt{2\pi}}\, dx$$

*or, equivalently, that*

$$\lim_{n\to\infty} P\left(\overline{X}_n \in \left[\mu + a\sigma/\sqrt{n}, \mu + b\sigma/\sqrt{n}\right]\right) = \int_a^b \frac{e^{-x^2/2}}{\sqrt{2\pi}}\, dx$$

$$= \int_{\mu+a\sigma/\sqrt{n}}^{\mu+b\sigma/\sqrt{n}} \frac{e^{-n(x-\mu)^2/(2\sigma^2)}}{\sigma\sqrt{2\pi}/\sqrt{n}}\, dx$$

and

$$\lim_{n\to\infty} P\left(Y_n \in [n\mu + a\sigma\sqrt{n}, n\mu + b\sigma\sqrt{n}]\right) = \int_a^b \frac{e^{-x^2/2}}{\sqrt{2\pi}}\, dx$$

$$= \int_{n\mu+a\sigma\sqrt{n}}^{n\mu+b\sigma\sqrt{n}} \frac{e^{-(x-n\mu)^2/(2n\sigma^2)}}{\sigma\sqrt{2\pi}\sqrt{n}}\, dx.$$

Note: for a proof of the Central Limit Theorem the reader is referred to [A], Section 8.3.

**3.4.2 Example.** Assume that $(X_k)_{k\in\mathbb{N}}$ is a sequence of discrete i.i.d. random variables, with common density

$$f(x) := \begin{cases} 1/3 & \text{if } x = 1, \\ 1/6 & \text{if } x = 2, \\ 1/2 & \text{if } x = 4, \\ 0 & \text{otherwise.} \end{cases}$$

Given this definition, the common mean and variance of the random variables $X_k$ are

$$\mu = 1 \cdot \frac{1}{3} + 2 \cdot \frac{1}{6} + 4 \cdot \frac{1}{2} = \frac{8}{3},$$

$$\sigma^2 = 1^2 \cdot \frac{1}{3} + 2^2 \cdot \frac{1}{6} + 4^2 \cdot \frac{1}{2} - \left(\frac{8}{3}\right)^2 = \frac{17}{9}.$$

Thus the Central Limit Theorem allows us to infer that for all $a, b \in \mathbb{R}$ with $a < b$ and all sufficiently large $n$ we have

$$P\left(\frac{8n}{3} + \frac{a\sqrt{17n}}{3} \le \sum_{k=1}^{n} X_k \le \frac{8n}{3} + \frac{b\sqrt{17n}}{3}\right) \approx \int_a^b \frac{e^{-x^2/2}}{\sqrt{2\pi}}\, dx.$$

We could easily add many other examples of this kind but we would probably gain little insight by doing so. For the fact is simply, as we said before, that the Central Limit Theorem *always* applies whenever trials are performed independently in an identical physical fashion. Whether the distribution of the random variables that represent these trials is binomial, or geometric, or something else—it doesn't matter—the Central Limit Theorem can be applied. There are no exceptions.

That said, there still may occur to the reader the following question: how do we find the values for $\mu$ and $\sigma$ that are so prominently used in the Central

Limit Theorem in a case where the distribution of the random variables $X_k$ is not known and cannot be inferred *a priori*? If a coin is known to be perfectly unbiased, then we also know that the distribution of each $X_k$ is binomial with parameters $n = 1$ and $p = 1/2$, but unfortunately, no coin is ever perfectly unbiased and its $p$-value therefore is never exactly $1/2$. So how large is the bias and how do we apply the Central Limit Theorem if we don't know $\mu$ and $\sigma$ because we don't know $p$? To provide an answer to these and some related questions is the purpose of the next and final chapter of this text.

## Exercises

**3.4.3.** A die is rolled 400 times. Use a standard normal integral to find an estimate for the probability that the average of the 400 values between 1 and 6 thus produced differs from the expected value 3.5 by no more than 0.1.

**3.4.4.** Assume that $X_1, X_2, \ldots, X_{900}$ are independent geometric random variables with the common geometric/binomial parameter $p = 1/4$. Use the Central Limit Theorem to estimate the probability for the average

$$\frac{1}{900} \sum_{k=1}^{900} X_k$$

to be greater than or equal to 4.2.

**3.4.5.** Assume that $X_1, X_2, \ldots, X_{400}$ are independent Poisson random variables with the common sample-size and density parameters $r = 5$ and $\delta = 2$, respectively. Use the Central Limit Theorem to estimate the probability for the average

$$\frac{1}{400} \sum_{k=1}^{400} X_k$$

to be contained in the interval $[9.7, 10.1]$.

# Chapter 4

# Estimation

## 4.1 Unbiased Estimators

If a random process of independent but physically identical trials produces a *sample* of values $x_1, \ldots, x_n$, it is natural to ask how the common mean $\mu$ and variance $\sigma^2$ of the underlying i.i.d. random variables $X_1, \ldots, X_n$ can be *estimated* from this sample. Considering first the mean, the simple answer is that $\mu$ is most adequately estimated by the *sample average* or *sample mean*

$$\overline{x}_n := \frac{1}{n} \sum_{k=1}^{n} x_k.$$

To justify this claim, we observe that the expected value of the random variable

$$\overline{X}_n := \frac{1}{n} \sum_{k=1}^{n} X_k$$

is

$$E(\overline{X}_n) = \frac{1}{n} \sum_{k=1}^{n} E(X_k) = \frac{1}{n} \sum_{k=1}^{n} \mu = \mu.$$

That is to say, the average $\overline{x}_n$ is an *unbiased* estimator for $\mu$ because the expectation of the corresponding random variable $\overline{X}_n$ is equal to $\mu$. So, in general, when we speak of an unbiased estimator, this is what we mean:

> An estimator of a certain quantity is *unbiased* if the expected value of the corresponding random variable is equal to that quantity.

Turning to the variance $\sigma^2 = E((X - E(X))^2)$, it is obviously tempting to think that an unbiased estimator for $\sigma^2$ might be the average

$$\frac{1}{n}\sum_{k=1}^{n}(x_k - \overline{x}_n)^2,$$

but unfortunately, the expected value of the corresponding random variable

$$\frac{1}{n}\sum_{k=1}^{n}(X_k - \overline{X}_n)^2$$

is not equal to $\sigma^2$—it is almost equal to $\sigma^2$ but not exactly. As it turns out, the proper unbiased estimator here is the so-called *sample variance*

$$\boxed{s_n^2 := \frac{1}{n-1}\sum_{k=1}^{n}(x_k - \overline{x}_n)^2}$$

because, according to Theorem 4.1.2 below, the expected value of the random variable

$$\boxed{S_n^2 := \frac{1}{n-1}\sum_{k=1}^{n}(X_k - \overline{X}_n)^2}$$

is equal to $\sigma^2$, as it ought to be.

Furthermore, if we are given additional random variables $Y_1, \ldots, Y_n$ such that $X_i$ and $Y_j$ are independent whenever $i \neq j$ and such that the joint distributions of the pairs $(X_1, Y_1), \ldots, (X_n, Y_n)$ are identical, then all these pairs have the same covariance, and an unbiased estimator for this covariance is the *sample covariance*

$$\boxed{q_n := \frac{1}{n-1}\sum_{k=1}^{n}(x_k - \overline{x}_n)(y_k - \overline{y}_n)}$$

because the expected value of the random variable

$$\boxed{Q_n := \frac{1}{n-1}\sum_{k=1}^{n}(X_k - \overline{X}_n)(Y_k - \overline{Y}_n)}$$

happens to be equal to precisely this covariance (see Theorem 4.1.4).

**4.1.1 Theorem.** *For any random variables* $X_1, \ldots, X_n, Y_1, \ldots, Y_n : S \to \mathbb{R}$ *it is the case that*

$$S_n^2 = \frac{1}{n-1}\left(\sum_{k=1}^{n}X_k^2 - n\overline{X}_n^2\right)$$

and

$$Q_n = \frac{1}{n-1} \left( \sum_{k=1}^{n} X_k Y_k - n \overline{X}_n \overline{Y}_n \right).$$

*Proof.*

$$S_n^2 = \frac{1}{n-1} \sum_{k=1}^{n} (X_k - \overline{X}_n)^2 = \frac{1}{n-1} \sum_{k=1}^{n} (X_k^2 - 2X_k \overline{X}_n + \overline{X}_n^2)$$

$$= \frac{1}{n-1} \left( \sum_{k=1}^{n} X_k^2 - 2\overline{X}_n \sum_{k=1}^{n} X_k + n\overline{X}_n^2 \right)$$

$$= \frac{1}{n-1} \left( \sum_{k=1}^{n} X_k^2 - 2n\overline{X}_n^2 + n\overline{X}_n^2 \right) = \frac{1}{n-1} \left( \sum_{k=1}^{n} X_k^2 - n\overline{X}_n^2 \right)$$

and, similarly,

$$Q_n = \frac{1}{n-1} \sum_{k=1}^{n} (X_k - \overline{X}_n)(X_k - \overline{X}_n)$$

$$= \frac{1}{n-1} \sum_{k=1}^{n} (X_k Y_k - X_k \overline{Y}_n - Y_k \overline{X}_n + \overline{X}_n \overline{Y}_n)$$

$$= \frac{1}{n-1} \left( \sum_{k=1}^{n} X_k Y_k - n \overline{X}_n \overline{Y}_n \right),$$

as desired. $\qquad \square$

**4.1.2 Theorem.** *If $X_1, \ldots, X_n : S \to \mathbb{R}$ are independent random variables with the common mean $\mu$ and common variance $\sigma^2$, then $E(S_n^2) = \sigma^2$.*

*Proof.* Using the representation in Theorem 4.1.1 in conjunction with Theorem 2.9.8, we find that

$$E(S_n^2) = \frac{1}{n-1} \left( \sum_{k=1}^{n} E(X_k^2) - nE\left( \left( \frac{1}{n} \sum_{k=1}^{n} X_k \right)^2 \right) \right)$$

$$= \frac{1}{n-1} \left( \sum_{k=1}^{n} (\sigma^2 + \mu^2) - \frac{n}{n^2} E\left( \left( \sum_{k=1}^{n} X_k \right)^2 \right) \right)$$

$$= \frac{n}{n-1} \left( \sigma^2 + \mu^2 - \frac{1}{n^2} E\left( \sum_{i,j=1}^{n} X_i X_j \right) \right)$$

$$= \frac{n}{n-1} \left( \sigma^2 + \mu^2 - \frac{1}{n^2} \left( \sum_{k=1}^{n} E(X_k^2) + \sum_{\substack{i,j=1 \\ j \neq i}}^{n} E(X_i X_j) \right) \right)$$

$$= \frac{n}{n-1} \left( \sigma^2 + \mu^2 - \frac{1}{n^2} \left( n(\sigma^2 + \mu^2) + \sum_{\substack{i,j=1 \\ j \neq i}}^{n} E(X_i)E(X_j) \right) \right)$$

$$= \frac{n}{n-1} \left( \left( 1 - \frac{1}{n} \right) (\sigma^2 + \mu^2) - \frac{1}{n^2} \sum_{\substack{i,j=1 \\ j \neq i}}^{n} \mu^2 \right)$$

$$= \sigma^2 + \mu^2 - \frac{n(n^2 - n)\mu^2}{(n-1)n^2} \quad \text{(see Exercise 4.1.8 below)}$$

$$= \sigma^2,$$

as desired. $\qquad\qquad\qquad\qquad\qquad\qquad\qquad\qquad\qquad\qquad\qquad\quad\square$

**4.1.3 Example.** The sample mean of the values $x_1 = 3$, $x_2 = 1$, $x_3 = 5$, $x_4 = 4$, and $x_5 = 2$ is

$$\overline{x}_5 = \frac{3+1+5+4+2}{5} = 3,$$

and, according to Theorem 4.1.1, the sample variance is

$$s_5^2 = \frac{1}{4} \left( 3^2 + 1^2 + 5^2 + 4^2 + 2^2 - 5 \cdot 3^2 \right) = \frac{5}{2}.$$

**4.1.4 Theorem.** *Assume that $X_1, \ldots, X_n, Y_1, \ldots, Y_n : S \to \mathbb{R}$ are random variables that satisfy following conditions: the pairs $(X_1, Y_1), \ldots, (X_n, Y_n)$ have the same joint density $f(x, y)$, and $X_i$ and $Y_j$ are independent whenever $i \neq j$. Then*

$$E(Q_n) = \mathrm{Cov}(X_1, Y_1) = \cdots = \mathrm{Cov}(X_n, Y_n).$$

*Proof.* The assumption that the pairs $(X_1, Y_1), \ldots, (X_n, Y_n)$ have a common joint density trivially implies that $E(X_1) = \cdots = E(X_n)$, $E(Y_1) = \cdots = E(Y_n)$, and $E(X_1 Y_1) = \cdots = E(X_n Y_n)$. Hence

$$\mathrm{Cov}(X_1, Y_1) = \cdots = \mathrm{Cov}(X_n, Y_n),$$

and therefore, we only need to show that $E(Q_n) = \mathrm{Cov}(X_1, Y_1)$. Using the representation of $Q_n$ in Theorem 4.1.1, we find that

$$E(Q_n) = \frac{1}{n-1} \left( \sum_{k=1}^{n} E(X_k Y_k) - nE(\overline{X}_n \overline{Y}_n) \right)$$

$$= \frac{1}{n-1}\left(nE(X_1Y_1) - nE(\overline{X}_n\overline{Y}_n)\right)$$

$$= \frac{1}{n-1}\left(nE(X_1Y_1) - \frac{1}{n}E\left(\left(\sum_{i=1}^{n}X_i\right)\left(\sum_{j=1}^{n}Y_j\right)\right)\right)$$

$$= \frac{1}{n-1}\left(nE(X_1Y_1) - \frac{1}{n}\sum_{i,j=1}^{n}E(X_iY_j)\right)$$

$$= \frac{1}{n-1}\left(nE(X_1Y_1) - \frac{1}{n}\sum_{i=1}^{n}E(X_iY_i) - \frac{1}{n}\sum_{\substack{i,j=1\\j\neq i}}^{n}E(X_i)E(Y_j)\right)$$

$$= \frac{1}{n-1}\left(nE(X_1Y_1) - \frac{1}{n}\sum_{i=1}^{n}E(X_1Y_1) - \frac{1}{n}\sum_{\substack{i,j=1\\j\neq i}}^{n}E(X_1)E(Y_1)\right)$$

$$= \frac{1}{n-1}\left((n-1)E(X_1Y_1) - \frac{(n^2-n)E(X_1)E(Y_1)}{n}\right)$$

$$= E(X_1Y_1) - E(X_1)E(Y_1)$$

$$= \text{Cov}(X_1,Y_1),$$

as desired. □

**4.1.5 Definition.** The *sample correlation coefficient* of two sets of sample values $M_x = \{x_1,\ldots,x_n\}$ and $M_y = \{y_1,\ldots,y_n\}$ with positive sample variances $s_n^2(x)$ and $s_n^2(y)$ is

$$r_n := \frac{q_n}{s_n(x)s_n(y)}.$$

Note: the values $s_n(x)$ and $s_n(y)$ are naturally referred to as the *sample standard deviations* of $M_x$ and $M_y$, respectively.

**4.1.6 Example.** Suppose that

$$x_1 = 1, \ x_2 = 2, \ x_3 = 2, \ x_4 = 4, \ x_5 = 6$$

and

$$y_1 = 3, \ y_2 = 6, \ y_3 = 5, \ y_4 = 2, \ y_5 = 9.$$

Then

$$\overline{x}_5 = \frac{1+2+2+4+6}{5} = \frac{15}{5} = 3,$$

$$\overline{y}_5 = \frac{3+6+5+2+9}{5} = \frac{25}{5} = 5,$$

$$s_5(x) = \sqrt{\frac{1^2 + 2^2 + 2^2 + 4^2 + 6^2 - 5 \cdot 3^2}{4}} = \sqrt{16/4} = 2,$$

$$s_5(y) = \sqrt{\frac{3^2 + 6^2 + 5^2 + 2^2 + 9^2 - 5 \cdot 5^2}{4}} = \sqrt{30/4} = \sqrt{7.5},$$

and

$$r_5 = \frac{1 \cdot 3 + 2 \cdot 6 + 2 \cdot 5 + 4 \cdot 2 + 6 \cdot 9 - 5 \cdot 3 \cdot 5}{2\sqrt{7.5} \cdot 4} = \frac{12}{8\sqrt{7.5}}$$

$$= \frac{3}{2\sqrt{7.5}} \approx 0.548.$$

To conclude this section, we will now demonstrate that, in essence, Theorem 2.11.10 is valid for *sample* correlation coefficients as well.

**4.1.7 Theorem.** *The sample correlation coefficient $r_n$ of two sets of sample values $\{x_1, \ldots, x_n\}$ and $\{y_1, \ldots, y_n\}$ with positive sample variances $s_n^2(x)$ and $s_n^2(y)$ always satisfies the inequality*

$$-1 \leq r_n \leq 1.$$

*Furthermore, $r_n = \pm 1$ if and only if there are constants $m, b \in \mathbb{R}$ such that $m \neq 1$ and $y_k = mx_k + b$ for all $k \in \{1, \ldots, n\}$. More precisely, in this latter case we have $r_n = 1$ if $m > 0$ and $r_n = -1$ if $m < 0$.*

*Proof.* In order to be able to apply Theorem 2.11.10, we set $X(k) := x_k$, $Y(k) := y_k$, and $P(\{k\}) := 1/n$ for all $k \in S := \{1, \ldots, n\}$. Then $E(X) = \bar{x}_n$, $E(Y) = \bar{y}_n$,

$$\mathrm{Var}(X) = \sum_{k=1}^{n} \frac{X(k)^2}{n} - \bar{x}_n^2 = \frac{1}{n}\left(\sum_{k=1}^{n} x_k^2 - n\bar{x}_n^2\right) = \frac{(n-1)s_n^2(x)}{n},$$

$$\mathrm{Var}(Y) = \frac{(n-1)s_n^2(y)}{n},$$

and

$$\mathrm{Cov}(X, Y) = \sum_{k=1}^{n} \frac{X(k)Y(k)}{n} - \bar{x}_n\bar{y}_n = \frac{1}{n}\left(\sum_{k=1}^{n} x_k y_k - n\bar{x}_n\bar{y}_n\right) = \frac{(n-1)q_n}{n}.$$

Consequently, Theorem 2.11.10 implies that

$$r_n = \frac{q_n}{s_n(x)s_n(y)} = \frac{(n-1)q_n/n}{\sqrt{(n-1)s_n^2(x)/n}\sqrt{(n-1)s_n^2(y)/n}} = \rho(X, Y) \in [-1.1].$$

Furthermore, given this fact that $r_n = \rho(X, Y)$, the remaining statements concerning the case where $r_n = \pm 1$ directly follow from the analogous statements in Theorem 2.11.10. $\qquad\square$

## Exercises

**4.1.8.** Explain why

$$\sum_{\substack{i,j=1 \\ j \neq i}}^{n} \mu^2 = (n^2 - n)\mu^2.$$

**4.1.9.** Find the sample mean and the sample variance of the values $x_1 = 1$, $x_2 = 2$, $x_3 = 5$, $x_4 = 2$, $x_5 = 6$, $x_6 = 1$, $x_7 = 3$, $x_8 = 2$, $x_9 = 4$, and $x_{10} = 2$.

**4.1.10.** Find the sample correlation coefficient $r_5$ for the following sets of data: $x_1 = 1$, $x_2 = 1$, $x_3 = 2$, $x_4 = 2$, $x_5 = 4$ and $y_1 = 1$, $y_2 = 3$, $y_3 = 3$, $y_4 = 4$, $y_5 = 4$.

**4.1.11.** Find the sample correlation coefficient $r_5$ for the following sets of data: $x_1 = 1$, $x_2 = 2$, $x_3 = 3$, $x_4 = 4$, $x_5 = 5$ and $y_1 = 3$, $y_2 = 5$, $y_3 = 7$, $y_4 = 9$, $y_5 = 11$. What does your answer tell you about the location of the points $(x_1, y_1)$, $(x_2, y_2)$, $(x_3, y_3)$, $(x_4, y_4)$, and $(x_5, y_5)$ in the $xy$-plane?

## 4.2    Regression Lines and Scatter Plots

Whenever empirical data is collected in a laboratory experiment, we need to choose adequate methods of data analysis and evaluation. Frequently useful in this context are visual representations in the form of graphs. For if we display, for example, a set of pairs of input-output data points $(x_1, y_1), \ldots, (x_n, y_n)$ in a two-dimensional *scatter plot*, we may be able, by looking at this plot, to discover a functional relation between the $x$- and $y$-values.

**4.2.1 Example.** Let us assume that an experiment, which measures the velocity of a falling body at different points in time, has generated the following results:

| $t$ in $s$ | 0 | 1 | 2 | 3 | 4 | 5 |
|---|---|---|---|---|---|---|
| $v$ in $m/s$ | 5.07 | 14.76 | 26.55 | 34.82 | 42.36 | 54.10 |

The corresponding scatter plot in a two-dimensional coordinate system (Figure 4.1) suggests the presence of a straight line dependence that ought to be modeled by an equation of the form

$$v = mt + b. \tag{4.1}$$

But how do we choose $m$ and $b$ in such a way that the straight line described by this equation approximates the points in Figure 4.1 as closely as possible? In order to answer this question, we need to introduce a quantitative measure for the overall distance between the line described by equation (4.1) and the points in Figure 4.1. And to this end we denote by $t_k$ and $v_k$ the values in the

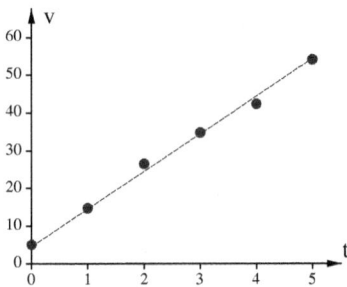

Figure 4.1: the scatter plot.

table above (i.e., $t_1 = 0, \ldots, t_6 = 5$ and $v_1 = 5.07, \ldots, v_6 = 54.10$) and define $D(m, b)$ to be the sum of the squares of the differences between the values $v_k$ and the $v$-coordinates $mt_k + b$ on the line:

$$D(m, b) := \sum_{k=1}^{6} (v_k - (mt_k + b))^2. \tag{4.2}$$

Given this definition, the problem of finding a best-fitting line for the points plotted in Figure 4.1 can be reformulated as an optimization problem in the sense that we need to find the values for $m$ and $b$ for which $D(m, b)$ is minimal.

To be sure, just as the definition of the variance, the definition of $D$ above is somewhat arbitrary (see the discussion on p.50). But here as there the square of a difference is easier to work with than, say, the absolute value of a difference, and it is also simpler and therefore more natural than, say, a difference raised to the power four. So the definition of $D$ in (4.2) is perfectly reasonable and also represents the conventional standard approach (notice, though, that a viable alternative will be offered in Exercise 4.2.12 at the end of this section). So in general, whenever we wish to determine a best fitting line—or *regression line*—for a given set of sample points $(x_1, y_1), \ldots, (x_n, y_n)$ (with $s_n(x), s_n(y) > 0$), we will minimize the function

$$D(m, b) := \sum_{k=1}^{6} (y_k - (mx_k + b))^2. \tag{4.3}$$

As a matter of course, optimization problems are well familiar from the calculus, and the first step here is always to find the critical points of the function that is to be optimized. More specifically, since $D$ depends on two variables $m$ and $b$, we need to find the points $(m, b)$ where the gradient of $D$ vanishes, that is, the points where

$$\nabla D(m, b) = \begin{pmatrix} \partial D(m, b)/\partial m \\ \partial D(m, b)/\partial b \end{pmatrix} = \begin{pmatrix} 0 \\ 0 \end{pmatrix}$$

Using elementary rules of differentiation, we find that

$$\nabla D(m,b) = \begin{pmatrix} \partial D/\partial m(m,b) \\ \partial D/\partial b(m,b) \end{pmatrix} = \begin{pmatrix} -\sum_{k=1}^{n} 2x_k(y_k - (mx_k + b)) \\ -\sum_{k=1}^{n} 2(y_k - (mx_k + b)) \end{pmatrix}, \quad (4.4)$$

and therefore, we need to solve for $m$ and $b$ the equations

$$\sum_{k=1}^{n} x_k(y_k - (mx_k + b)) = 0,$$

$$\sum_{k=1}^{n} (y_k - (mx_k + b)) = 0.$$

The first of these equations yields

$$\sum_{k=1}^{n} x_k y_k = \sum_{k=1}^{n} x_k(mx_k + b) = \left(\sum_{k=1}^{n} x_k^2\right) m + n\bar{x}_n b, \quad (4.5)$$

and the second equation implies that

$$n\bar{y}_n = \sum_{k=1}^{n} y_k = \sum_{k=1}^{n} (mx_k + b) = n\bar{x}_n m + nb. \quad (4.6)$$

In solving the system of linear equations represented by (4.5) and (4.6) for $m$ and $b$, we easily obtain

$$m = \frac{\sum_{k=1}^{n} x_k y_k - n\bar{x}_n \bar{y}_n}{\sum_{k=1}^{n} x_k^2 - n\bar{x}_n^2} = \frac{q_n}{s_n^2(x)} = \frac{s_n(y)}{s_n(x)} r_n \quad (4.7)$$

and

$$b = \frac{\bar{y}_n \sum_{k=1}^{n} x_k^2 - \bar{x}_n \sum_{k=1}^{n} x_k y_k}{\sum_{k=1}^{n} x_k^2 - n\bar{x}_n^2}$$

$$= \frac{\bar{y}_n \left(\sum_{k=1}^{n} x_k^2 - n\bar{x}_n^2\right) - \bar{x}_n \left(\sum_{k=1}^{n} x_k y_k - n\bar{x}_n \bar{y}_n\right)}{\sum_{k=1}^{n} x_k^2 - n\bar{x}_n^2} \quad (4.8)$$

$$= \bar{y}_n - \frac{s_n(y)}{s_n(x)} r_n \bar{x}_n.$$

Consequently, the equation of the best fitting regression line is

$$y = \frac{s_n(y)}{s_n(x)} r_n x + \bar{y}_n - \frac{s_n(y)}{s_n(x)} r_n \bar{x}_n$$

or, equivalently,

$$\boxed{\frac{(y - \bar{y}_n)/s_n(y)}{(x - \bar{x}_n)/s_n(x)} = r_n.} \quad (4.9)$$

In words: the regression line through the points $(x_1, y_1), \ldots, (x_n, y_n)$ consist of precisely those points $(x, y)$ for which the ratio of the *sample z-scores*

$$\frac{y - \bar{y}_n}{s_n(y)} \quad \text{over} \quad \frac{x - \bar{x}_n}{s_n(x)}$$

is equal to the sample correlation coefficient $r_n$. Put differently, if we place the origin of a coordinate system at $(\bar{x}_n, \bar{y}_n)$ and scale the axes so that one unit on the $x$-axis equals $s_n(x)$ and one unit on the $y$-axis equals $s_n(y)$, then, in this shifted and rescaled coordinated system, the regression line passes through the origin with slope $r_n$.

*Remark.* If we wish, we can apply the second derivative test (as stated, for instance, in [B2], p.707) to verify that $D$ assumes a local minimum at the critical point given by (4.7) and (4.8). Using (4.4), it is easy to see that

$$\frac{\partial^2}{\partial m^2} D(m, b) = \sum_{k=1}^{n} 2x_k^2, \quad \frac{\partial^2}{\partial b^2} D(m, b) = 2n, \quad \text{and} \quad \frac{\partial^2}{\partial b \partial m} D(m, b) = \sum_{k=1}^{n} 2x_k.$$

Hence the discriminant is

$$4n \sum_{k=1}^{n} x_k^2 - 4 \left( \sum_{k=1}^{n} x_k \right)^2 = 4n \left( \sum_{k=1}^{n} x_k^2 - n\bar{x}_n^2 \right) = 4n(n-1)s_n^2(x)$$

and the last term on the right is obviously positive whenever $n > 1$ (which we may assume because finding a regression line is not possible if only one point is given). Thus—by the second derivative test—the critical point given by (4.7) and (4.8) is not a saddle point but rather a local extremum. But since $\partial^2 D(m, b)/\partial b^2 = 2n > 0$, we may further infer that this local extremum is indeed a minimum.

**4.2.2 Example.** For the values $v_k$ and $t_k$ given in Example 4.2.1 we have

$$\bar{t}_6 = \frac{1}{6} \sum_{k=1}^{6} t_k = \frac{0 + 1 + 2 + 3 + 4 + 5}{6} = \frac{15}{6},$$

$$\bar{v}_6 = \frac{1}{6} \sum_{k=1}^{6} v_k = \frac{5.07 + 14.76 + 26.55 + 34.82 + 42.36 + 54.10}{6} = \frac{177.66}{6},$$

and

$$\sum_{k=1}^{6} t_k^2 = 55,$$

$$\sum_{k=1}^{6} t_k v_k = 612.26.$$

Substituting for the corresponding expressions in (4.7) and (4.8) (with $t_k$ and $v_k$ in place of $x_k$ and $y_k$, respectively) yields

$$m = \frac{612.26 - 15 \cdot 177.66/6}{55 - 15^2/6} \approx 9.61$$

and

$$b = \frac{55 \cdot 177.66/6 - 15 \cdot 612.26/6}{55 - 15^2/6} \approx 5.59.$$

These results show that the data in Example 4.2.1 describe the motion of a falling body with an approximate initial velocity of $5.59\,m/s$ and an approximately constant acceleration of $9.61\,m/s^2$. So relative to the actual acceleration $g = 9.81\,m/s^2$ we have an error of about 2%.

**4.2.3 Example.** In Section 2.11 we introduced the functions

$$X(t) = \cos(2t)$$

and

$$Y_1(t) = 1.1\cos(2t - 0.3),$$
$$Y_2(t) = 1.1\cos(2t - 1.4),$$
$$Y_3(t) = 1.1\cos(2t - 2.8)$$

as highly simplified hypothetical models for the dependence of atmospheric temperatures and $CO_2$ concentrations on time. In reality, of course, empirical scientific research does not generate smooth functional dependencies but rather sets of individual measurement values. For instance, if we choose to perform measurements at points in time that are multiples of, say, $2\pi k/11$, then the functions $X$ and $Y_1$ produce the following table of values:

| $t$ | $X(t)$ | $Y_1(t)$ | $t$ | $X(t)$ | $Y_1(t)$ |
|---|---|---|---|---|---|
| 0 | 1 | 1.05 | $12\pi/11$ | 0.84 | 1.06 |
| $2\pi/11$ | 0.42 | 0.73 | $14\pi/11$ | −0.14 | 0.17 |
| $4\pi/11$ | −0.65 | −0.44 | $16\pi/11$ | −0.96 | −0.92 |
| $6\pi/11$ | −0.96 | −1.1 | $18\pi/11$ | −0.65 | −0.93 |
| $8\pi/11$ | −0.14 | −0.47 | $20\pi/11$ | 0.42 | 0.14 |
| $10\pi/11$ | 0.84 | 0.71 | $22\pi/11$ | 1 | 1.05 |

Naturally, the values in the $X$- and $Y_1$-columns may be paired up so as to provide us with the following list of $xy$-data points:

$$(x_1, y_1) = (1, 1.05)$$
$$(x_2, y_2) = (0.42, 0.73)$$
$$\vdots$$
$$(x_{12}, y_{12}) = (1, 1.05).$$

In order to determine the regression line that these points specify, we observe that

$$\overline{x}_{12} = \frac{1 + 0.42 + \cdots + 1}{12} = 0.085,$$

$$\overline{y}_{12} = \frac{1.05 + 0.73 + \cdots + 1.05}{12} = 0.0875,$$

and

$$\sum_{k=1}^{12} x_k^2 = 6.4914,$$

$$\sum_{k=1}^{12} x_k y_k = 6.8239.$$

According to (4.7) and (4.8), this yields

$$m = \frac{6.8239 - 12 \cdot 0.085 \cdot 0.0875}{6.4914 - 12 \cdot 0.085^2} \approx 1.052$$

and

$$b = \frac{0.0875 \cdot 6.4914 - 0.085 \cdot 6.8239}{6.4914 - 12 \cdot 0.085^2} \approx -0.002,$$

and the corresponding scatter plot of the data points $(x_1, y_1), \ldots, (x_{12}, y_{12})$ with the regression line $y = 1.052x - 0.002$ is shown in Figure 4.2. Given

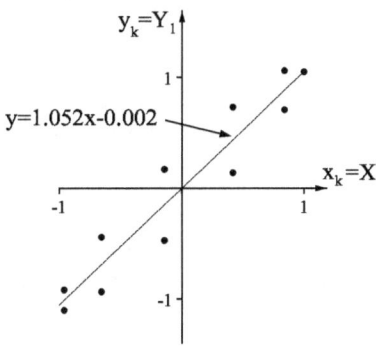

Figure 4.2: scatter plot for $(x_k, y_k) = (X, Y_1)$.

this diagram, we may recall that in Figure 2.18 the slope of the limiting lines above and below the curve $c_1 = (X, Y_1)$ was almost the same as the slope of the regression line in Figure 4.2—1.051 instead of 1.052. Moreover, the $y$-intercept $b = -0.002$ of the regression line is very nearly zero. So perhaps these slight deviations from the values 1.051 and 0 are due to the fact that we only considered finitely many values in the computation of $m$ and $b$. Perhaps a more accurate computation that takes into account the full information provided by the function $X$ and $Y_1$ would reveal that the 'ideal regression line' is

$$y = 1.051x + 0 = 1.051x. \tag{4.10}$$

In other words, what we are here proposing is a generalization of the methods discussed in this section that allows us to find a best fitting line for a continuous curve—in this case the curve that is described by the parameterization $c_1(t) = (X(t), Y_1(t))$ for $t \in S = [0, 2\pi]$. To sort this matter out, we replace the finite-case regression line formula (4.9) by the following, very natural continuous case analogue:

$$\boxed{\frac{(y - E(Y))/\sigma(Y)}{(x - E(X))/\sigma(X)} = \rho(X, Y).} \tag{4.11}$$

According to (2.51) and Example 2.11.7, we have $E(X) = E(Y_1) = 0$, $Var(X) = 1/2$, $Var(Y_1) = 0.605$, and

$$E(XY_1) = \frac{1.1 \cos(0.3)}{2} \approx 0.525.$$

Consequently, (4.11) assumes in this present example the following form:

$$\frac{y/\sqrt{0.605}}{x/\sqrt{1/2}} = \frac{1.1 \cos(0.3)/2}{\sqrt{1/2}\sqrt{0.605}}$$

or, equivalently,

$$y = 1.1 \cos(0.3)x \approx 1.051x, \tag{4.12}$$

as conjectured. Moreover and to be sure, the fact that the slope values in (4.10) and (4.12) appear to be equal is not due to rounding because a comparison with (2.55) shows that the exact slope in both equations is $m = 1.1 \cos(0.3)$.

## Exercises

**4.2.4.** Find the equation of the regression line that best fits the points $(0, 1)$, $(2, 3)$, $(5, 8)$, and $(6, 9)$, and plot these points together with the line on the same set of axes.

**4.2.5.** Find the regression line for the pair $(X_\alpha, Y_\alpha)$ for all $\alpha \in [0, 2\pi]$, where

$$X_\alpha(s) := 3 \cos(\alpha) \cos(s) - \sin(\alpha) \sin(s)$$

and

$$Y_\alpha(s) := 3 \sin(\alpha) \cos(s) + \cos(\alpha) \sin(s),$$

for all $s \in [0.2\pi]$ and where the density on $S := [0, 2\pi]$ is assumed to be uniform.

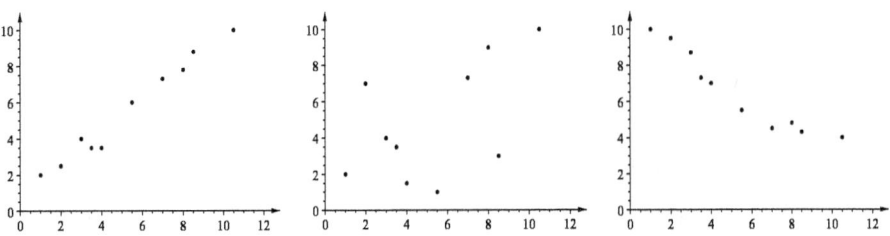

Figure 4.3: scatter plots.

**4.2.6.** For which of the scatter plots shown in the diagrams in Figure 4.3 is the sample correlation coefficient largest and for which is it closest to zero?

**4.2.7.** What will happen if the point $A$ is removed from the scatter plot in Figure 4.4 below? Will the sample correlation coefficient increase or will it decrease? Explain your answer.

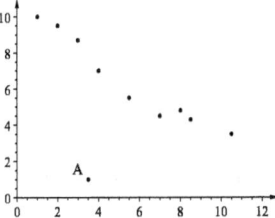

Figure 4.4: scatter plot with point $A$.

**4.2.8.** What will happen if the point $A$ is removed from the scatter plot in Figure 4.5 below? Will the sample correlation coefficient increase or will it decrease? Explain your answer.

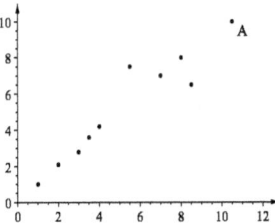

Figure 4.5: scatter plot with point $A$.

**4.2.9.** Use (4.9) to show that the regression line that best approximates two points $(x_1, y_1)$ and $(x_2, y_2)$ with $x_1 \neq x_2$ and $y_1 \neq y_2$ is precisely the line that passes through these two points.

**4.2.10.** Let $X(s) := s$ and $Y_n(s) := s^n$ for all $s \in S := [0, 1]$ (where the density on $S$ is uniform). Find the regression line for the curve that is described by the parameterization $c_n(s) := (X(s), Y_n(s))$. Furthermore, show that the limiting regression line, as $n$ tends to infinity, is $y = 0$ and explain why this observation is intuitively not surprising.

**4.2.11.** Use the same method as in the case of linear regression to determine a *parabola*, described by an equation of the form $y = ax^2 + bx + c$, that best fits the points $(-1, 1)$, $(1, -1)$, $(2, 0)$, and $(5, 6)$. Then plot these points and the parabola on the same set of axes.

**4.2.12.** In the definition of $D$ in (4.3) we measured the distance between a point $(x_k, y_k)$ and the line $y = mx + b$ by taking the square of the difference of the $y$-coordinates (i.e., $(mx_k + b - y_k)^2$). But the actual least distance is not this difference in the $y$-coordinates (or the absolute value of this difference) but rather the absolute value of this difference divided by $\sqrt{1 + m^2}$ (see, for instance, [B1], Chapter 12). So the square of this least distance is

$$\frac{(mx_k + b - y_k)^2}{1 + m^2}.$$

Furthermore, if we pass again from a finite set of points $(x_k, y_k)$ to a continuous curve $c(s) = (X(s), Y(s))$, defined on some continuous sample space $S$, as we did in Example 4.2.3, then the square of the least distance between $(X(s), Y(s))$ and the line $y = mx + b$ is

$$\frac{(mX(s) + b - Y(s))^2}{1 + m^2},$$

and therefore, the overall distance measure that needs to be minimized is

$$D(m, b) = \int_S \frac{(mX(s) + b - Y(s))^2}{1 + m^2} p(s)ds.$$

So if we choose for simplicity's sake $S = [0, 1]$ and $p(s) = 1$, then

$$D(m, b) = \int_0^1 \frac{(mX(s) + b - Y(s))^2}{1 + m^2} ds.$$

Show that in using this approach the slope of the regression line turns out to be

$$m = \frac{\text{Var}(Y) - \text{Var}(X)}{2\,\text{Cov}(X, Y)} \pm \sqrt{\left(\frac{\text{Var}(Y) - \text{Var}(X)}{2\,\text{Cov}(X, Y)}\right)^2 + 1}$$

(whenever $\text{Cov}(X, Y) \neq 0$) and that the equation of the regression line is

$$\frac{y - E(Y)}{x - E(X)} = m.$$

## 4.3   Randomness and Consciousness

Depending on our personal inclinations and needs, we may rightly consider probability to be either a self-contained mathematical theory—an end in itself—or, alternatively, a rational tool for exploring and understanding external reality. Adopting here the latter point of view, the question that arises is to what extent we are actually willing to let our beliefs concerning the nature of that reality be molded or defined by the hard-core quantitative information that probabilistic analyses supposedly generate. Should we consider the conclusions drawn from such analyses to be objective factual statements that cannot be questioned or should we rather trust them only if they tend to reinforce our pre-conceived notions of how the world works? The answer is far from obvious, and the philosophical treatises that have been written in trying to find it are very likely very numerous—and voluminous.

However, since there is obviously no room in a mathematical textbook for far flung epistemological excursions, we will here put aside the deeper questions of philosophy and direct our attention instead at an actual real-world example. That is to say, what we will do is to discuss a seemingly rigorous statistical study, conducted at Princeton University, that reportedly produced strongly positive results but that nonetheless was highly controversial because it was highly worldview-sensitive.

The Princeton Engineering Anomalies Research (PEAR) program was established in the spring of 1979 and remained in operation for twenty-eight years until February 2007. In order to understand its purpose and founding idea, it is helpful to look at some prior developments. The field of parapsychology, to which the PEAR program loosely belonged, traces its origins, as an academic discipline, to the late 1920s when J.B. and Louisa Rhine joined the Duke University Department of Psychology to study such purported psychic phenomena as extrasensory perception and telekinesis. In some of their typical experiments the Rhines asked selected individuals to identify sequences of geometric symbols on randomized cards or to influence mentally the roll of a die (see [JD2], pp.43–4, and [McT], p.101, for more details). In assessing their work many years later, J.B. Rhine concluded that the outcomes obtained had been unmistakably positive:

> The phenomena that were being studied began to show lawful interrelations and even a degree of unity. One by one the major claims, based originally only on spontaneous human experiences, were subjected to laboratory test and experimentally verified.[I]

With respect to the Rhines' experiments with dice we may say that one of these "major claims" is the very strange assertion that a person's willful intent

---

[I]This quote is taken from an address to the American Psychological Association by J.B. Rhine, in 1967, and can be found in [JD2], p.44.

can have a statistically significant effect on how a die falls. Consequently, it is advisable for us to try to understand to begin with, exactly how we can detect statistical significance in random experiments. Here is an example: a die is rolled 100 times and the number of times that it shows a 1, a 2, or a 3 is recorded. If the die is perfectly unbiased, the probability for a single roll to produce a 1, a 2, or a 3 is $1/2$, and the recorded outcome after 100 rolls therefore follows a binomial distribution with parameters $p = 1/2$ and $n = 100$. Thus, the expected value is

$$np = 100 \cdot \frac{1}{2} = 50,$$

and the standard deviation is

$$\sqrt{np(1-p)} = \sqrt{100 \cdot \left(\frac{1}{2}\right)^2} = 5.$$

Given these values, it wouldn't be surprising in the least if the recorded value after 100 trials turned out to be 52. For the deviation from the mean would here be $52 - 50 = 2$ which is less than one standard deviation. By contrast, if the recorded value after 10000 trials was 5200, we ought to be very surprised indeed because in this latter case the expected value is $10000/2 = 5000$, and the standard deviation is $\sqrt{10000 \cdot (1/2)^2} = 50$. So the deviation from the mean would be equal to 4 standard deviations (because $5200 - 5000 = 200 = 4 \cdot 50$) and would therefore be unusually large (by the 68-95-99.7 rule because a binomial distribution is approximately normal when $n$ is large).

In order to understand more fully the meaning of this example, it is helpful to represent each roll of the die by a binomial random variable $X_{n,k}$ with

$$P(X_{n,k} = 1) = \frac{1}{2} = P(X_{n,k} = 0),$$

(see Example 2.9.9). Then the values 52 and 5200, referred to above, are output values of the sums

$$\sum_{k=1}^{100} X_{100,k}$$

and

$$\sum_{k=1}^{10000} X_{10000,k},$$

respectively, and the value $52/100 = 5200/10000 = 0.52$ is a *common* output value of both

$$\overline{X}_{100} = \frac{1}{100} \sum_{k=1}^{100} X_{100,k}$$

and

$$\overline{X}_{10000} = \frac{1}{10000} \sum_{k=1}^{10000} X_{10000,k}.$$

Seen in this light, the contrast between our lack of surprise at receiving the value 52 after 100 trials and our intense surprise at receiving the value 5200 after 10000 trials is directly expressive of the far more general, underlying fact that, according to the weak law of large numbers (Theorem 3.2.4), it is the case that

$$0 = \lim_{n\to\infty} P\left(|\overline{X}_n - 0.5| > 0.02\right) = \lim_{n\to\infty} P\left(|\overline{X}_n - \mu| > 0.52 - \mu\right).$$

Furthermore, since this limit equation is valid not only for $\varepsilon := 0.02$ but for any $\varepsilon > 0$, it follows that any bias, no matter how slight, becomes detectable with arbitrarily high probability as the number of trials $n$ increases to infinity. So if indeed the willful intent of a conscious observer can somehow affect the roll of a die—very slightly presumably—then that minute influence will, in all likelihood, become apparent if only the number of rolls is sufficiently large. We may have to perform millions or billions of trials, but eventually and most likely, the influence will be detected.

For instance, if the shift in probability away from the ideal mean 0.5 is not 0.02 but only $0.0001 = 10^{-4}$, then the biased value for $p$ is 0.5001, and the expected value after one billion trials is

$$0.5001 \cdot 1,000,000,000 = 500,100,000.$$

Since the corresponding standard deviation is

$$\sqrt{1,000,000,000 \cdot 0.5001 \cdot 0.4999} \approx 15,811 \approx 16,000,$$

the chance that the actual value after ten billion trials is somewhere between $500,068,000$ and $500,132,000$ is about 95% (by the 68-95-99.7 rule). If we are very unlucky, the actual value may be as low as, say, $500,052,000$, but even then the presence of bias is still very obvious. For in absence of bias the expected value would be

$$1,000,000,000 \cdot 0.5 = 500,000,000,$$

and the deviation from this ideal mean would be

$$500,052,000 - 500,000,000 = 52,000,$$

that is, it would be equal to more than 3 standard deviations and thus very unlikely. Furthermore, if we increase the number of trials from one billion to ten billion, say, then the difference in the expected values is

$$10,000,000,000 \cdot 0.5001 - 10,000,000,000 \cdot 0.5 = 1,000,000$$

and the standard deviation without bias is

$$\sqrt{10,000,000,000 \cdot 0.5^2} = 50,000.$$

Consequently, the difference between the ideal and the biased mean is equal to 20 standard deviations and thus extremely large. So absent a cosmic coincidence, we will be able to correctly conclude—after ten billion trials—that there does exist a very slight bias.

That said, however, there still remains a problem. For the amount of time needed for millions or billions of rolls of a die is obviously enormous. Moreover, the Rhines' experimental protocol was not only slow but also lacking in rigor. One obvious flaw was due to the fact that no die can ever be assumed to be perfectly regular and thus perfectly unbiased, and another arose from the possibility of errors being made in the counting of outcomes by hand. The resulting data distortions were likely minute, but nonetheless potentially significant as the influence of conscious intent was naturally thought to be very slight as well.

A partial solution to these problems was found by a theoretical physicist, named Helmut Schmidt, who had been inspired by the Rhines' research to conduct his own telekinetic experiments. Schmidt began his pertinent studies as a research fellow at Boeing, in the mid nineteen-sixties, and then joined the Rhines' laboratory at Duke, in 1970. The central thrust of his work was to improve the Rhines' protocol by objectifying the underlying random process. Instead of a die rolled by a human operator he employed a random number generator (RNG) that utilized the inherent quantum-mechanical randomness of radioactive decay. Furthermore, to run his experiments, Schmidt enlisted the help of prominent psychics whom he asked to either predict or consciously influence the output that his RNGs yielded (for more information see [S1], [S2], and [McT], pp.101–9). Unfortunately, though, the rate of data production, even in Schmidt's improved experiments, was still fairly low, and his reliance on psychics could be criticized as being bizarre.

Consequently, further improvements were needed, and this is where our story returns to the program at Princeton and here in particular to a gifted young sophomore in the school of engineering who had recently chanced upon some of Schmidt's papers. Feeling deeply intrigued by the adventurous notion that human consciousness could perhaps directly affect a machine, this student approached Robert Jahn, then dean of engineering at Princeton, with a proposal to independently verify the validity of Schmidt's empirical claims. After some initial hesitation, Jahn agreed to supervise a two-year undergraduate project but only on condition that the student would first conduct a thorough bibliographical search and manage to convince him that the topic at hand had sufficient provisional credibility to warrant its further pursuit. When the student completed this task, to Jahn's satisfaction, she was given permission to proceed and began to develop her own experimental designs. As the project progressed and was starting to yield some compelling results, Jahn found himself getting ever more deeply involved, making suggestions and giving advice (see [McT], p.110). At the end of two years of fruitful exploration the student concluded that the phenomena under investigation were real. But soon thereafter she went on to graduate and hence abandoned her psychic research. In consequence, Jahn was left with a set of empirical data that had not only piqued his curiosity but also given him a growing sense of concern that modern technology's increasingly sensitive information processing systems might be vulnerable to the "effects suggested by [his student's] pilot experiments" ([JD2], p.89). And thus he decided to found the engineering anomalies research program at Princeton.

The primary goal of this program was to make more rigorous the prior work of Schmidt and the Rhines. One of the most prominent experiments that Jahn and his main collaborator, Brenda Dunne, conducted involved a type of RNG that may be best described as a microelectronic coin-tossing device.[IV] This machine randomly produced zeros and ones at such a high a rate that more data could be amassed in the PEAR laboratory during a single afternoon than the Rhines had collected in a lifetime (see [McT], p.113). And it was this vastly superior rate of data acquisition that made possible a far more discerning statistical analysis—as compared to the work of Schmidt and the Rhines—and also made unnecessary Schmidt's controversial reliance on psychics. That is to say, the operators chosen by Jahn and Dunne were ordinary men and women without any claims to unusual paranormal abilities.

According to the standard PEAR protocol, each operator was placed in front of an RNG machine and was instructed to attempt to affect the machine's output in accordance with three different forms of pre-recorded volition: increasing the number of ones (positive intent), increasing the number of zeros (negative intent), or leaving the output unchanged (neutral intent). Supposedly, these choices of intent really did make a difference. Again and again it was found that small shifts in output occurred as operators changed their stated intent. Moreover and very interestingly, the individual operators were found to have distinct effectual signatures: some achieved large shifts, others only small ones; some exerted symmetric effects,[V] others did not; for some the direction of the shift was opposite to their intent, for others it was in agreement; some achieved better results when choosing their intent freely, others did better when being instructed (see [JD2], pp.104–15 and also [JDN]). In fact, the operators' typical signatures appeared to be even dependent on gender,[VI] and pairs of operators could exert compound effects that differed

---

[IV] To be precise, Jahn and Dunne referred to these machines as 'random event generators' and used the acronym REG rather than RNG (see [JD2], pp.91–103, for more information).

[V] That is to say, the effects for positive and negative intents were approximately equal.

[VI] According to the findings in [D1], the effects exerted by males are more strongly correlated with pre-recorded intentions and also show a smaller statistical spread.

considerably from those produced by the operators separately.[VII] And finally, the supposedly neutral intent, which frequently *did* leave the mean output invariant, turned out to be not entirely neutral after all. For the corresponding data sets commonly had a reduced statistical spread—a smaller sample variance. In other words, the neutral intent somehow had a constraining influence in that it lowered the output values' natural propensity to fluctuate (see [D2] for more details).

It goes without saying that most self-respecting scientists will relegate such extravagant claims to the realm of delusion. Reality is material, consciousness is epiphenomenal, and hence there cannot be any connection between subjective states of intent and strictly external physical systems. The notion is clearly ridiculous. But then again the PEAR research was not produced by imbeciles, and the data that Robert Jahn and his collaborators collected must somehow be accounted for. If the data is sound, the implications are profound. But is it sound and are we willing to accept it?

We cannot here pursue these questions any further, and we also cannot survey exhaustively all the various results and experiments that the PEAR team produced and conducted. In other words, it's up to the reader to decide whether he or she wishes to study this subject more deeply (see the expositions in [JD1] and [JD2] and also in Chapter 5 of [B3] for more information). That said, we now will direct our attention at some of the statistical methods and tools that the PEAR team employed (in somewhat modified form to suit the present exposition). For these are prototypical and highly instructive and therefore ideally suited for an introductory textbook.

Consider for instance the simulated graph in Figure 4.6 which shows the output of a hypothetical RNG experiment. For simplicity we have here chosen to model, not the actual tripolar protocol used in the PEAR lab, but rather a bipolar protocol with positive and negative intent but no neutral intent. So what this graph shows is how much the high and low intent curves, in the course of our hypothetical experiment, deviated from the expected mean, as given by the horizontal axis, and what it shows as well is a horizontal parabola that represents a deviation from this mean in the positive and negative directions by one standard deviation. In order to properly calculate this standard deviation, we need to understand that one unit on the horizontal axis represents one thousand trials with 200 electronic coin flips each. (Note: the latter number of 200 per basic run of the RNG device is consistent with the actual design of the PEAR experiments.) So one unit actually represents 200,000 flips. Consequently, if we denote the horizontal axis variable by $m$, then the number of electronic coin flips corresponding to $m$ is $200000m$, and the standard

---

[VII]Here again, the operator's gender did play a role. Opposite-sex pairs achieved greater effects than same-sex pairs, and in the former group the effects were greatest for pairs with deep emotional bonds (see [D2]).

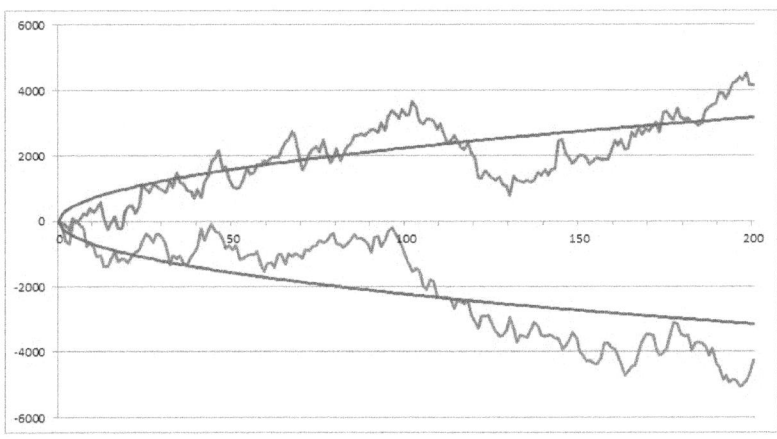

Figure 4.6: a simulated PEAR graph with high and low intent curves.

deviation therefore is

$$\sqrt{200000m \cdot \left(\frac{1}{2}\right)^2} = 100\sqrt{5m}.$$

In other words, if we denote by $d$ (for 'deviation') the dependent variable displayed on the vertical axis, then the equation for the horizontal parabola in Figure 4.6 is

$$d = \pm 100\sqrt{5m}.$$

The fact that both the high and the low intent curves differ from the mean after $200 \cdot 200,000 = 40,000,000$ trials by a bit more than one standard deviation (see Figure 4.6) is not all that remarkable. But insofar as our purpose is to determine the conjectured influence of conscious intent, we may also examine (as an added example), not the absolute deviation of each curve, but rather the relative deviation of the two curves from each other. Hence we are shown in Figure 4.7 the (added) graph of the difference of the high and low intent curves together with a horizontal semi-parabola that represents the position of the 95th percentile. But what exactly does this mean? To answer this question, we will denote by $X_m$ and $Y_m$ the binomial random variables that represent, respectively, the high and low intent outcomes after $200000m$ trials. If the RNG that produces the data was properly calibrated and if there was no influence upon it by conscious intent, then the binomial parameters of both $X_m$ and $Y_m$ would be $n = 200000m$ and $p = 1/2$. Consequently, $X_m$ and $Y_m$ would be approximately normal with mean

$$\mu_m = \frac{200000m}{2} = 100000m$$

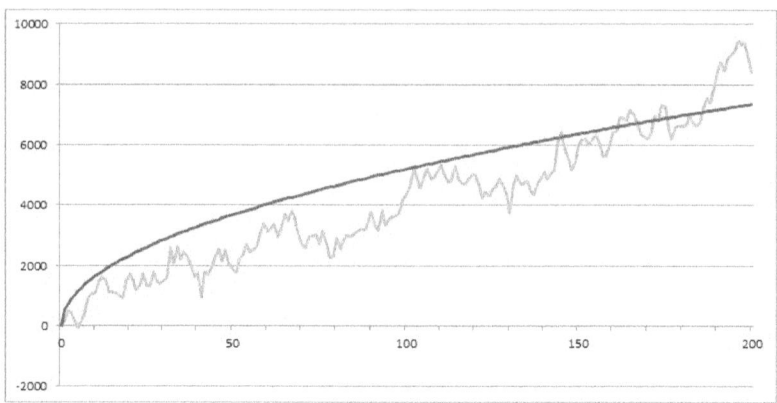

Figure 4.7: the difference between the high and low intent curves.

and standard deviation

$$\sigma_m = 100\sqrt{5m}.$$

Since $X_m$ and $Y_m$ can obviously be assumed to be independent (because the electronic coin flips are presumably independent of each other), we may apply Theorem 2.9.12 to infer that the random variable $X_m - Y_m$ is approximately normal with mean

$$\mu_m - \mu_m = 0$$

and standard deviation

$$\sqrt{\sigma_m^2 + \sigma_m^2} = \sqrt{2}\sigma_m.$$

By implication, the approximate density of $X_m - Y_m$ is

$$f_m(x) = \frac{e^{-x^2/(4\sigma_m^2)}}{\sqrt{2\pi}\sqrt{2}\sigma_m},$$

and the vertical $d$-coordinate of the 95th-percentile semi-parabola in Figure 4.7 is approximately given by the equation

$$0.95 = P(X_m - Y_m \le d) = \int_{-\infty}^{d} f_m(x)\,dx = \int_{-\infty}^{d/(\sqrt{2}\sigma_m)} \frac{e^{-x^2/2}}{\sqrt{2\pi}}\,dx.$$

Using a computer to solve this equation numerically for $d/(\sqrt{2}\sigma_m)$ yields

$$1.64485 \approx \frac{d}{\sqrt{2}\sigma_m},$$

or equivalently,

$$d \approx 1.64485\sqrt{2}\sigma_m = 164.485\sqrt{10m}.$$

So this is the defining equation of the semi-parabola shown in Figure 4.7, and the significance of this parabola is that the chance for the graph of $X_m - Y_m$ to be above this parabola at any point $m$ is always (approximately) equal to 5%. Since the graph in Figure 4.7 ends up above the semi-parabola at $m = 200$, it follows that there is a less than 5% chance for the high intent curve to be as far above the low intent curve as it actually is in Figure 4.6.

Another representational tool that we find in PEAR reports is the scatter plot with a best-fitting normal density curve. To see what this is about, we will consider as an example the following table of values (which again is not actual but artificially generated):

| heads | high | low | heads | high | low | heads | high | low |
|-------|------|-----|-------|------|-----|-------|------|-----|
| 72 | 0 | 1 | 91 | 268 | 245 | 110 | 207 | 175 |
| 73 | 0 | 0 | 92 | 294 | 258 | 111 | 169 | 167 |
| 74 | 1 | 6 | 93 | 346 | 359 | 112 | 114 | 152 |
| 75 | 1 | 0 | 94 | 371 | 427 | 113 | 127 | 122 |
| 76 | 0 | 0 | 95 | 439 | 430 | 114 | 50 | 67 |
| 77 | 2 | 14 | 96 | 451 | 450 | 115 | 62 | 40 |
| 78 | 2 | 1 | 97 | 495 | 537 | 116 | 64 | 33 |
| 79 | 4 | 3 | 98 | 552 | 555 | 117 | 43 | 49 |
| 80 | 19 | 31 | 99 | 588 | 558 | 118 | 4 | 11 |
| 81 | 3 | 8 | 100 | 561 | 560 | 119 | 27 | 2 |
| 82 | 20 | 14 | 101 | 587 | 555 | 120 | 22 | 10 |
| 83 | 8 | 51 | 102 | 530 | 509 | 121 | 1 | 7 |
| 84 | 62 | 68 | 103 | 498 | 513 | 122 | 10 | 1 |
| 85 | 40 | 50 | 104 | 485 | 499 | 123 | 0 | 0 |
| 86 | 95 | 52 | 105 | 411 | 459 | 124 | 2 | 1 |
| 87 | 75 | 109 | 106 | 406 | 381 | 125 | 5 | 0 |
| 88 | 101 | 162 | 107 | 359 | 313 | 126 | 0 | 0 |
| 89 | 185 | 170 | 108 | 340 | 337 | 127 | 0 | 0 |
| 90 | 224 | 238 | 109 | 269 | 240 | 128 | 1 | 0 |

What this table shows is the output data of a hypothetical bipolar-protocol experiment in which an RNG produces two sets of electronic coin flips while supposedly being influenced by an operator's high and low intent. As before, each run of the RNG generates 200 hundred coin flips and the number of runs is 10000 for both of these pre-recorded intents. Looking at the range displayed in the 'head'-columns, we see that the smallest number of heads that the RNG produced in a total of $2 \cdot 10000 = 20000$ runs was 72 and the largest was 128. Furthermore, each entry in either the high or the low intent columns tells us how many times the corresponding number in the 'head'-column was produced. For instance, there were 42 high-intent and 48 low-intent runs that produced 84 heads (and 116 tails) because the value 84 in the first 'head'-column is lined up with the values 42 and 48 in the first high-intent and the first low-intent column, respectively.

Consistent with this description, we find that the values in the three high-intent columns add up to 10000 and so do the values in the three low intent columns, that is,

$$0 + 0 + 1 + 1 + 0 + 2 + \cdots + 0 + 2 + 5 + 0 + 0 + 1 = 10000$$

and

$$1 + 0 + 1 + 3 + 2 + 4 + \cdots + 3 + 1 + 0 + 0 + 0 + 0 = 10000.$$

Using a computer, we readily find, by implication, that the average number of heads is

$$\bar{x}_{10000}^{+} = \frac{0 \cdot 72 + 0 \cdot 73 + 1 \cdot 74 + 1 \cdot 75 + 0 \cdot 76 + \cdots + 1 \cdot 128}{10000} = 100.1896$$

for the high intent mode and

$$\bar{x}_{10000}^{-} = \frac{1 \cdot 72 + 0 \cdot 73 + 1 \cdot 74 + 3 \cdot 75 + 2 \cdot 76 + \cdots + 0 \cdot 128}{10000} = 99.8029$$

for the low intent mode. Moreover, the corresponding sample variances are

$$(s_{10000}^{+})^2 = \frac{0 \cdot (72 - \bar{x}_{10000}^{+})^2 + \cdots + 1 \cdot (128 - \bar{x}_{10000}^{+})^2}{9999} \approx 49.6782$$

and

$$(s_{10000}^{-})^2 = \frac{1 \cdot (72 - \bar{x}_{10000}^{+})^2 + \cdots + 0 \cdot (128 - \bar{x}_{10000}^{+})^2}{9999} \approx 50.2307.$$

Consequently, the sample standard deviations are

$$s_{10000}^{+} \approx \sqrt{49.4558} \approx 7.0483$$

and

$$s_{10000}^{-} \approx \sqrt{50.0197} \approx 7.0874,$$

respectively. Given these values, the best-fitting normal density functions—rescaled to 100 percent—are

$$f^{+}(x) = \frac{100e^{-(x - \bar{x}_{10000}^{+})^2 / 2(s_{10000}^{+})^2}}{\sqrt{2\pi} s_{10000}^{+}}$$

and

$$f^{-}(x) = \frac{100e^{-(x - \bar{x}_{10000}^{-})^2 / 2(s_{10000}^{-})^2}}{\sqrt{2\pi} s_{10000}^{-}}.$$

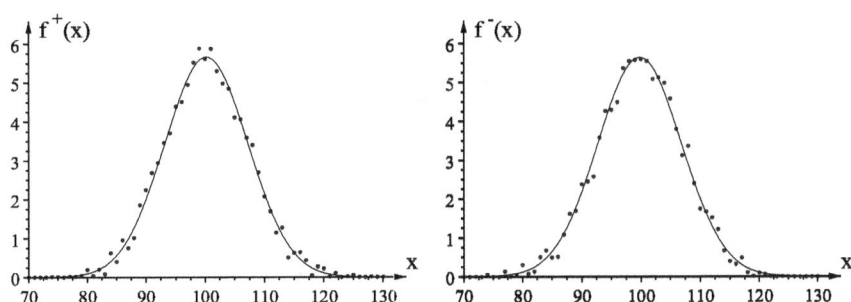

Figure 4.8: high and low intent scatter plots with normal densities.

Finally, as we now rescale the values from the high and low columns in the table above to 100 hundred percent as well (by multiplying each value by $100/10000 = 0.01$), we can display these values in two scatter plots together with the graphs of $f^+$ and $f^-$, respectively (see Figure 4.8). In addition we can also plot the approximating normal graphs together in the same coordinate system so as to better see the shifting effect of the two types of intent. The resulting image (in Figure 4.9) appears to show only a very slight discrepancy

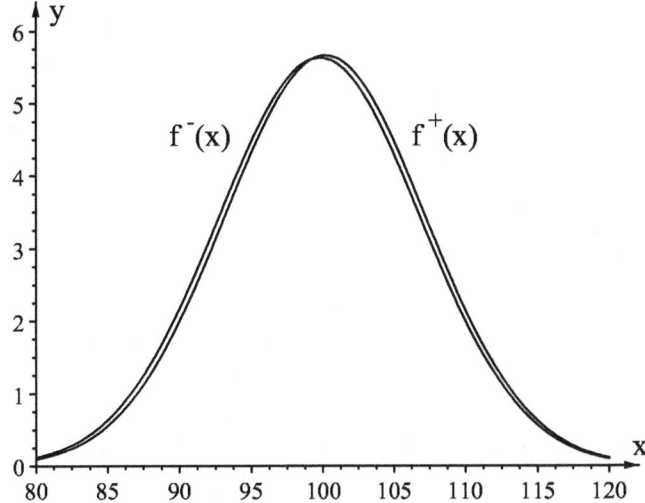

Figure 4.9: high and low intent normal graphs.

between the graphs of $f^+$ and $f^-$, but here plain eyesight is misleading. For if there was no operator influence, then the averages $\overline{x}^+_{10000}$ and $\overline{x}^-_{10000}$ would be output values of a binomial random variable with parameters $n = 10000 \cdot 200 = 2,000,000$ and $p = 1/2$ divided by 10000, and the standard deviation of this random variable would be equal to

$$\sqrt{\frac{2000000(1/2)^2}{10000^2}} \approx 0.0707.$$

Consequently, the difference

$$\overline{x}^+_{10000} - \overline{x}^-_{10000} = 100.1896 - 99.8029 = 0.3867$$

would be larger than 5 standard deviations and thus very significant.

In summary and in conclusion we wish to point out that the complexity and variety of the statistical methods used in PEAR reports are greater than the few examples that we here chose to discuss are able to convey. But given the overall purpose and scope of the present exposition, the selection seems adequate. For the chosen examples nicely demonstrate how a number of core concepts that we previously introduced—in Chapters 2 and 3 as well as in Section 4.1—can be fruitfully employed in a very concrete application context.

## Exercises

**4.3.1.** Assume that a random process produces heads and tails (or ones and zeros). Whenever the process has been run 200 times, we count the number of heads (or ones) and record it as the outcome of a random experiment. The table below shows (in the 'trials'-columns) how many times the possible head counts from 0 to 200 have occurred in a total of 10000 trials. This means that the numbers in all the 'trials'-columns add up to 10000, and it also means, for example, that the bottommost entries 64 and 41 in the first two columns tell us that 41 out of 10000 trials produced 64 heads in 200 electronic coin flips. Furthermore, there were no trials that produced less than 50 heads or more than 109 heads because these head counts are not listed in the table and are therefore understood to not have occurred.

a) Use an Excel worksheet to determine the sample average and the sample standard deviation of the head counts of the 10000 trials recorded in the table.

b) Use a) to find a normal density function that approximately models the distribution of the i.i.d. random variables $X_k$ that may be thought to be associated with the 10000 experiments that the table pertains to.

c) Plot the density function that you found in c) rescaled in such a way that the $y$-axis represents a chance percentage.

d) Use the results of a) to find a normal density function that approximately models the distribution of the random variable $X := \sum_{k=1}^{10000} X_k$.

e) Use the density that you found in d) to estimate the probability that $X$ differs from its expected value by more than 1000.

**f)** Use the density that you found in d) to find an approximation for a value $z$ such that $P(X \le E(X) - z) = 0.05$.

**g)** Would you be surprised to learn that ideally each random variable $X_k$ has a binomial distribution with parameters $p = 0.4$ and $n = 200$?

**h)** Would you be surprised to learn that ideally each random variable $X_k$ has a binomial distribution with parameters $p = 0.4015$ and $n = 200$?

| heads | trials | heads | trials | heads | trials | heads | trials |
|-------|--------|-------|--------|-------|--------|-------|--------|
| 50 | 0 | 65 | 50 | 80 | 577 | 95 | 58 |
| 51 | 0 | 66 | 73 | 81 | 567 | 96 | 41 |
| 52 | 0 | 67 | 101 | 82 | 551 | 97 | 27 |
| 53 | 1 | 68 | 130 | 83 | 524 | 98 | 18 |
| 54 | 0 | 69 | 161 | 84 | 487 | 99 | 17 |
| 55 | 0 | 70 | 209 | 85 | 439 | 100 | 11 |
| 56 | 1 | 71 | 253 | 86 | 392 | 101 | 5 |
| 57 | 3 | 72 | 295 | 87 | 338 | 102 | 2 |
| 58 | 2 | 73 | 350 | 88 | 293 | 103 | 4 |
| 59 | 4 | 74 | 401 | 89 | 246 | 104 | 1 |
| 60 | 9 | 75 | 448 | 90 | 198 | 105 | 0 |
| 61 | 14 | 76 | 494 | 91 | 168 | 106 | 0 |
| 62 | 16 | 77 | 519 | 92 | 129 | 107 | 0 |
| 63 | 30 | 78 | 551 | 93 | 98 | 108 | 0 |
| 64 | 41 | 79 | 576 | 94 | 77 | 109 | 0 |

# 4.4 Confidence Intervals

The two parameters that we have encountered again and again, throughout this text, are the mean or expectation $\mu$ and the variance $\sigma^2$. In Section 4.1 we learned that unbiased estimators for these two parameters are the sample average $\bar{x}_n$ and the sample variance $s_n^2$, respectively, and in Section 4.3 we saw how these estimators can be fruitfully employed for instance in determining a best-fitting normal density curve. That said, it is natural for us to further inquire how well $\mu$ and $\sigma^2$ are actually approximated by $\bar{x}_n$ and $s_n^2$, statistically speaking. That is to say, what we should ask, more precisely, is how probable it is for $\mu$ and $\sigma^2$ to be contained in a certain interval $[a, b]$ that represents to us a certain statistical spread or margin of error. In order to properly address this question, it is helpful to introduce the following terminology:

> For a given probability value $q \in (0, 1)$, we say that an interval $[a, b]$ is a $100q\%$ *confidence interval* for $\mu$ (or $\sigma^2$) if the probability for $\mu$ (or $\sigma^2$) to be contained in $[a, b]$ is equal to $q$.

For the sake of simplicity, we will consider first a couple of examples in which a confidence interval for $\mu$ is determined under the assumption that the variance is known. This is a bit artificial and not fully satisfactory, but estimating the variance is rather involved and will therefore be dealt with separately in Section 4.5.

**4.4.1 Example.** Given a *normal* random variable $X : S \to \mathbb{R}$, we wish to find a 95% confidence interval $[a, b]$ for the mean $\mu = E(X)$ under the assumption that $\sigma^2 = 4 = Var(X)$. What does this mean and how would we do this in practice? First of all it means that $X$ represents some kind of a physical random process with a normally distributed numerical output (like the $x$-coordinate of a point where a dart hits a plane) the expected value of which we do not know. In order to estimate that expected value, practically speaking, we need to perform that random process repeatedly so as to create a sample of output values $x_1, \ldots, x_n$ and then take the corresponding sample average $\overline{x}_n$. For it is this sample average that represents the desired estimate of the expected value $\mu$. Furthermore, since a normal density is alway symmetric relative to $\mu$, it is natural for us to determine a 95% confidence interval that is symmetric relative to $\overline{x}_n$. To do so, we need to find an $\varepsilon > 0$ such that the chance for the unknown mean $\mu$ to be contained in the interval

$$[a, b] = [\overline{x}_n - \varepsilon, \overline{x}_n + \varepsilon]$$

is 95%. In other words, the probability for the inequalities

$$\overline{x}_n - \varepsilon \leq \mu \leq \overline{x}_n + \varepsilon$$

to be satisfied ought to be 0.95. But in order for this to make any sense, we need to interpret the values $x_1, \ldots, x_n$ to be output values of random variables $X_1, \ldots, X_n : S \to \mathbb{R}$, because otherwise there simply is no probability or chance for us to talk about. Moreover, since the values $x_1, \ldots, x_n$ are produced independently by the same random process, it follows that the random variables $X_1, \ldots, X_n$ must be construed to be independent and that the common distribution of them must be equal to the normal distribution of $X$. In other words and in particular, $X_1, \ldots, X_n$ are i.i.d. normal random variables with common mean $\mu$ and common variance $\sigma^2 = 4$. By implication then, what we are really looking for is a value $\varepsilon > 0$ so that $q = 0.95$ is the probability of the event $E \subset S$ that consists of all elements $s \in S$ for which it is the case that

$$\overline{X}_n(s) - \varepsilon \leq \mu \leq \overline{X}_n(s) + \varepsilon.$$

Since these two inequalities together are equivalent to the single inequality

$$|\overline{X}_n(s) - \mu| \leq \varepsilon,$$

it follows that $\varepsilon$ must satisfy the equation

$$q = 0.95 = P(|\overline{X}_n - \mu| \leq \varepsilon). \tag{4.13}$$

Since $\overline{X}_n$ is normal with mean $\mu$ and variance $\sigma^2/n = 4/n$ (by Theorem 2.9.12 and Lemma 3.2.5), it follows that

$$\frac{\overline{X}_n - \mu}{\sigma/\sqrt{n}}$$

is standard normal (by Corollary 2.6.7) and that, by implication,

$$q = 0.95 = P\left(\left|\frac{\overline{X}_n - \mu}{\sigma/\sqrt{n}}\right| \leq \frac{\varepsilon}{\sigma/\sqrt{n}}\right) = \int_{-\varepsilon\sqrt{n}/\sigma}^{\varepsilon\sqrt{n}/\sigma} \frac{e^{-x^2/2}}{\sqrt{2\pi}}\, dx$$

$$= 1 - 2\int_{-\infty}^{-\varepsilon\sqrt{n}/\sigma} \frac{e^{-x^2/2}}{\sqrt{2\pi}}\, dx$$

(4.14)

or, equivalently,

$$\frac{1-q}{2} = \frac{1}{40} = \int_{-\infty}^{-\varepsilon\sqrt{n}/\sigma} \frac{e^{-x^2/2}}{\sqrt{2\pi}}\, dx.$$

Using a computer to solve for $y$ the equation

$$\frac{1}{40} = \int_{-\infty}^{-y} \frac{e^{-x^2/2}}{\sqrt{2\pi}}\, dx,$$

we find that

$$\frac{\varepsilon\sqrt{n}}{\sigma} = y \approx 1.96,$$

and therefore,

$$\varepsilon \approx \frac{1.96\sigma}{\sqrt{n}}.$$

(4.15)

Consequently, if we perform for example $n = 100$ trials, then the 95% confidence interval for $\mu$ is

$$\left[\overline{x}_{100} - \frac{1.96 \cdot 2}{10}, \overline{x}_{100} + \frac{1.96 \cdot 2}{10}\right] = [\overline{x}_{100} - 0.392, \overline{x}_{100} + 0.392],$$

(4.16)

and if we increase the number of trials to, say, $n = 10,000$, then the 95% confidence interval is

$$[\overline{x}_{10000} - 0.0392, \overline{x}_{10000} + 0.0392].$$

(4.17)

Finally, in order for the average $\overline{x}_{100}$ or $\overline{x}_{10000}$ to be replaced by an actual value, we would have to actually perform 100 or 10,000 random trials and take the average of the outcomes that these trials produce. For example, if this average turned out to be 3.5156 after 10,000 trials, then the corresponding 95% confidence interval for $\mu$ would be

$$[3.5156 - 0.0392, 3.5156 + 0.0392] = [3.4764, 3.5548].$$

**4.4.2 Example.** If we drop the assumption of normality in the preceding example and only assume that $X : S \to \mathbb{R}$ is a random variable with $\mathrm{Var}(X) = \sigma^2 = 4$, then we have two basic options: we can use the Central Limit Theorem to find an approximate 95% confidence interval or we can use Chebyshev's inequality to find a somewhat larger upper-estimate interval. In the former case, the exact equation (4.14) is replaced by the approximate equation

$$q = 0.95 \approx 1 - 2 \int_{-\infty}^{-\varepsilon\sqrt{n}/\sigma} \frac{e^{-x^2/2}}{\sqrt{2\pi}} \, dx,$$

but the resulting approximate equation (4.15) reamains unaltered. That is to say, we still have

$$\boxed{\varepsilon \approx \frac{1.96\sigma}{\sqrt{n}},} \tag{4.18}$$

but the approximation is less precise because in addition to the rounding of the factor 1.96 we also use, as we said, the Central Limit Theorem. Usually, though, when $n$ is large, this is not a problem, and therefore, all the conclusions that followed in Example 4.4.1 from the approximate equation (4.15) can be inferred in exactly the same way in the present case from the identical approximate equation (4.18).

Turning now to the second option, referred to above, it is natural to replace to begin with the equal sign in (4.13) with a less-than-or-equal sign so as to match the corresponding sign in Chebyshev's inequality:

$$q = 0.95 \le P(|\overline{X}_n - \mu| \le \varepsilon). \tag{4.19}$$

This makes very good sense, because if the probability on the right-hand side of (4.19) should turn out to be strictly greater than 0.95, then we are even more than 95% certain that $\mu$ does not differ from $\overline{X}_n$ by more than $\varepsilon$, and that's obviously desirable. Using Theorem 3.2.4, we may infer that inequality (4.19) is satisfied whenever

$$1 - \frac{4}{\varepsilon^2 n} = 1 - \frac{\sigma^2}{\varepsilon^2 n} \ge 0.95$$

or, equivalently, whenever

$$\varepsilon \ge \sqrt{\frac{80}{n}}.$$

Since the term on the right-hand side of this inequality converges to zero as $n$ tends to infinity, it follows that the size of the confidence interval can be chosen to be as small as we wish if only the number $n$ of the sample values $x_1, \ldots, x_n$ is sufficiently large. For instance if $n = 100$, as in Example 4.4.1, then there is a chance greater than or equal to 95% for $\mu$ to be contained in the interval

$$\left[\overline{x}_{100} - \sqrt{\frac{80}{100}}, \overline{x}_{100} + \sqrt{\frac{80}{100}}\right] \approx [\overline{x}_{100} - 0.8944, \overline{x}_{80} + 0.8944],$$

and if $n = 10,000$, then this same interval is reduced to

$$[\overline{x}_{10000} - 0.0894, \overline{x}_{10000} + 0.0894].$$

Both of these intervals, however, are more than twice as wide as the corresponding intervals in (4.16) and (4.17), and this illustrates nicely that the approximation obtained by using Chebyshev's inequality—while being fully rigorous—is typically less precise than the one derived from the Central Limit Theorem.

As we said above, the case discussed in the preceding two examples where the variance is known but the mean is not is somewhat artificial. After all, if our insight into the nature of a given random process is sufficiently acute for us to compute the variance $E(X^2) - E(X)^2$ of its associated random variable $X$, then very probably this insight will enable us to find the expectation $E(X)$ as well. Consequently it may seem that in all cases that can be of genuine practical interest to us a confidence interval for $\mu$ cannot be determined without determining first—or in parallel—a confidence interval for $\sigma^2$ or $\sigma$ by using the estimator $s_n^2$. But here we need to be careful because there is another option to consider, namely the case where $\sigma$ is a function of $\mu$. For if $\sigma = \sigma(\mu)$ and $q = 0.95$, then, according to (4.18), we have

$$\varepsilon \approx \frac{1.96\sigma(\mu)}{\sqrt{n}}, \tag{4.20}$$

and there is thus a chance of approximately 95% that the following inequalities are satisfied:

$$\overline{x}_n - \frac{1.96\sigma(\mu)}{\sqrt{n}} \leq \mu \leq \overline{x}_n + \frac{1.96\sigma(\mu)}{\sqrt{n}}. \tag{4.21}$$

Consequently, if we can solve for $\mu$ both of these inequalities, then a potentially asymmetric confidence interval for $\mu$ can thereby be determined (where 'asymmetric' means 'not centered at $\overline{x}_n$').

**4.4.3 Example.** If $X : S \to \mathbb{R}$ is a binomial random variable with parameters $n = 1$ and $p$, then $\mu = p$ and $\sigma^2 = p(1 - p)$. Hence

$$\sigma(\mu) = \sqrt{\mu(1 - \mu)},$$

and therefore, (4.21) implies that

$$\overline{x}_n - \frac{1.96\sqrt{\mu(1 - \mu)}}{\sqrt{n}} \leq \mu \leq \overline{x}_n + \frac{1.96\sqrt{\mu(1 - \mu)}}{\sqrt{n}}$$

or, equivalently, that

$$(\overline{x}_n - \mu)^2 \leq \frac{1.96^2 \mu(1-\mu)}{n}.$$

Using the quadratic formula to solve this inequality for $\mu$, we readily find that

$$\boxed{\alpha_n - \beta_n \leq \mu \leq \alpha_n + \beta_n,}$$

where

$$\alpha_n := \frac{2\overline{x}_n + \dfrac{1.96^2}{n}}{2\left(1 + \dfrac{1.96^2}{n}\right)}$$

and

$$\beta_n := \frac{\sqrt{\dfrac{1.96^2}{n}\left(4\overline{x}_n(1-\overline{x}_n) + \dfrac{1.96^2}{n}\right)}}{2\left(1 + \dfrac{1.96^2}{n}\right)}.$$

Consequently, the 95% confidence interval for $\mu$ is

$$[\alpha_n - \beta_n, \alpha_n + \beta_n]. \tag{4.22}$$

As it turns out, this result can be fruitfully employed in opinion-poll sampling. To see how it works, let's suppose that we wish to find out how the people in a given population feel about a certain current-day issue or event. If the population is large, which it commonly is, then it very likely is far too costly and also too labor intensive to ask every single individual in that population for his or her opinion. So what we would do in a case like this is to try to select a random sample of individuals in such a way that the probability to be chosen for that sample is the same for all individuals. That is to say, we would randomly select individuals $I_1, \ldots, I_n$ (not necessarily all distinct), ask them each a yes-or-no question and assign the value 1 to a yes-answer and the value 0 to a no-answer. Proceeding in this way, we would effectively represent the answer given by an individual $I_k$ by the output value $x_k \in \{0, 1\}$ of a binomial random variable $X_k$ with parameters 1 and $p$, where $100p\%$ is the percentage of people in the overall population that would answer "yes" if they were asked. In other words, $p = \mu$ is the parameter that we wish to estimate by means of our random sample. So let's suppose for instance that the size of our sample is 900 and that 463 people in that sample vote "yes," then

$$\overline{x}_n = \overline{x}_{900} = \frac{463}{900},$$

and, by implication,

$$\alpha_n = \alpha_{900} = \frac{\dfrac{2 \cdot 463}{900} + \dfrac{1.96^2}{900}}{2\left(1 + \dfrac{1.96^2}{900}\right)} \approx 0.51438$$

and

$$\beta_n = \beta_{900} = \frac{\sqrt{\dfrac{1.96^2}{900}\left(\dfrac{4 \cdot 463}{900}\left(1 - \dfrac{463}{900}\right) + \dfrac{1.96^2}{900}\right)}}{2\left(1 + \dfrac{1.96^2}{900}\right)} \approx 0.03258.$$

Hence the 95% confidence interval for the actual *population proportion* $p = \mu$ is

$$[\alpha_{900} - \beta_{900}, \alpha_{900} + \beta_{900}] \approx [0.48180, 0.54697], \qquad (4.23)$$

and the corresponding 95% confidence interval for the actual population percentage $100p\%$ is

$$[48.180\%, 54.697\%].$$

Another less rigorous method that is frequently used in opinion poll sampling is to replace $\mu$ in (4.20) by its unbiased estimator $\bar{x}_n$. This yields

$$\boxed{\varepsilon \approx \frac{1.96\sqrt{\mu(1-\mu)}}{\sqrt{n}} \approx \frac{1.96\sqrt{\bar{x}_n(1-\bar{x}_n)}}{\sqrt{n}},} \qquad (4.24)$$

and using the same values as above, we find that

$$\varepsilon \approx \frac{1.96\sqrt{463/900(1 - 463/900)}}{\sqrt{900}} = \frac{1.96\sqrt{463 \cdot 437}}{27000} \approx 0.03265.$$

Thus the 95% confidence interval for $p$ here turns out to be

$$\left[\frac{463}{900} - 0.03265, \frac{463}{900} - 0.03265\right] \approx [0.48179, 0.54709],$$

which is slightly wider and therefore less precise than the interval (4.23). Clearly, this second method is not fully rigorous because if $\varepsilon$ represents the 95% margin of error in the value of $\bar{x}_n$ and if we use $\bar{x}_n$ to compute $\varepsilon$, then this margin of error should obviously also be taken into account in our computation of $\varepsilon$ but it is not. Fortunately, though, there is a way to sidestep this difficulty if we are willing to further reduce the precision of our estimate. All we need to do is to observe that the maximal value of the function $\sigma(\mu)$ (over the interval $[0, 1]$) is $1/2$ and that therefore

$$\boxed{\varepsilon \approx \frac{1.96\sqrt{\mu(1-\mu)}}{\sqrt{n}} \leq \frac{0.98}{\sqrt{n}}.} \qquad (4.25)$$

Using this upper estimate for $\varepsilon$—which no longer involves $\bar{x}_n$—with $n = 900$, we find that

$$\varepsilon \leq \frac{0.98}{30} \approx 0.03267.$$

So in this case, the rigorous upper estimate $0.03267$ is only very slightly larger than the value $\varepsilon \approx 0.03265$ that we found previously. As a note of caution, however, we wish to add that even this 'rigorous' upper estimate is of course not fully rigorous because underlying everything there is still the uncertainty induced by applying the Central Limit Theorem in our derivation of (4.20) (and (4.21) thereby).

**4.4.4 Example.** Assume that $X : S \to \mathbb{R}$ is a geometric random variable with mean $\mu = 1/p$ and variance $\sigma^2 = (1-p)/p^2$ (see Examples 2.3.2 and 2.3.9). Then

$$\sigma = \sigma(\mu) = \sqrt{\mu(\mu-1)},$$

and therefore,

$$\bar{x}_n - \frac{1.96\sqrt{\mu(\mu-1)}}{\sqrt{n}} \leq \mu \leq \bar{x}_n + \frac{1.96\sqrt{\mu(\mu-1)}}{\sqrt{n}}$$

or, equivalently,

$$(\bar{x}_n - \mu)^2 \leq \frac{1.96^2\mu(\mu-1)}{n}.$$

Using the quadratic formula, as we did above, in Example 4.4.3, yields

$$\gamma_n - \delta_n \leq \mu \leq \gamma_n + \delta_n,$$

where

$$\gamma_n := \frac{2\bar{x}_n - \dfrac{1.96^2}{n}}{2\left(1 - \dfrac{1.96^2}{n}\right)}$$

and

$$\delta_n := \frac{\sqrt{\dfrac{1.96^2}{n}\left(4\bar{x}_n(\bar{x}_n - 1) + \dfrac{1.96^2}{n}\right)}}{2\left(1 - \dfrac{1.96^2}{n}\right)}.$$

Thus, the 95% confidence interval for $\mu$ is

$$[\gamma_n - \delta_n, \gamma_n + \delta_n].$$

In all the preceding examples the confidence-interval probability was $q = 0.95$. If we change this value to, say, 0.9 or 0.99, we need to adjust our equations accordingly, and insofar as we are using a normal approximation via the Central Limit Theorem (or an actual normal density as in Example 4.4.1), the crucial step in this respect is to solve for $y = \varepsilon\sqrt{n}/\sigma$ the equation

$$\boxed{\frac{1-q}{2} = \int_{-\infty}^{-y} \frac{e^{-x^2/2}}{\sqrt{2\pi}}\, dx.} \qquad (4.26)$$

For $q = 0.95$ we found the solution $y \approx 1.96$, and for $q = 0.9$ and $q = 0.99$ the solutions are

$$\frac{\varepsilon\sqrt{n}}{\sigma} = y \approx 1.2816$$

and

$$\frac{\varepsilon\sqrt{n}}{\sigma} = y \approx 2.3263,$$

respectively. So in all places where we used previously the value 1.96 we would use instead the value 1.2816 or 2.3263 if we wish to compute a 90% or 99% confidence interval. It's really very simple.

## Exercises

**4.4.5.** Assume that $X$ is a normal random variable with unknown mean $\mu$ and known standard deviation $\sigma = 3$.

   a) Find a 90% confidence interval for $\mu$, given that $\overline{x}_{900} = 4.56$.

   b) Find a 95% confidence interval for $\mu$, given that $\overline{x}_{400} = 2.21$.

   c) Find a 97% confidence interval for $\mu$, given that $\overline{x}_{625} = 3.74$.

**4.4.6.** Use Chebyshev's inequality to find an upper estimate for a 97% confidence interval for $\mu = E(X)$, given that $\sigma^2 = \mathrm{Var}(X) = 25$ and $\overline{x}_{1600} = 25.45$.

**4.4.7.** An opinion poll is conducted by asking 1100 randomly selected adults a yes-or-no question. As it turns out, 609 adults answer 'yes' and 491 answer 'no'.

   a) Use (4.22) to find a 95% confidence interval for the percentage of 'yes'-answers.

   b) Use (4.24) to confirm your answer to a).

   c) Use (4.25) to confirm your answer to b).

**4.4.8.** An opinion poll is conducted by asking 1100 randomly selected adults a yes-or-no question. As it turns out, 192 adults answer 'yes' and 908 answer 'no'.

a) Use (4.22) to find a 95% confidence interval for the percentage of 'yes'-answers.

b) Use (4.24) to confirm your answer to a).

c) Try to confirm your answer to b) by means of (4.25). What do you notice and why?

**4.4.9.** Find a 95% confidence interval for the mean $\mu$ of a geometric random variable $X$ under the assumption that $\overline{x}_{900} = 7.43$.

**4.4.10.** Find a 90% confidence interval for the mean $\mu$ of a geometric random variable $X$ under the assumption that $\overline{x}_{1000} = 6.13$.

**4.4.11.** Let

$$\eta_n := \overline{x}_n + \frac{1.96^2}{2n}$$

for all $n \in \mathbb{N}$. Show that

$$\left[ \eta_n - \sqrt{\eta_n^2 - \overline{x}_n^2}, \eta_n + \sqrt{\eta_n^2 - \overline{x}_n^2} \right]$$

is an approximate 95% confidence interval for the mean $\mu$ of a Poisson random variable $X$. *Hint:* the mean of a Poisson random variable is equal to its variance (according to Exercise 2.4.2).

**4.4.12.** Find a 95% confidence interval for the mean of a Poisson random variable $X$ under the assumption that $\overline{x}_{2500} = 8.28$.

## 4.5 Estimating $\sigma^2$

As we now address the issue of finding confidence intervals for the variance $\sigma^2$, we need to review first a couple of basic facts concerning the gamma function (see [B1], Chapter 48, for more information).

**4.5.1 Definition.** The *gamma function* is defined via the equation

$$\Gamma(x) := \int_0^\infty t^{x-1} e^{-t} \, dt$$

for all $x > 0$.

**4.5.2 Theorem.** *For all $x > 0$ and all $n \in \mathbb{N} \cup \{0\}$ it is the case that*

**a)** $\Gamma(x+1) = x\Gamma(x)$ and

**b)** $\Gamma(n+1) = n!$.

*Proof.* **a)** Using integration by parts, we find that

$$\Gamma(x+1) = \int_0^\infty t^x e^{-t}\, dt = -t^x e^{-t}\Big|_0^\infty + \int_0^\infty x t^{x-1} e^{-t}\, dt = x\Gamma(x),$$

as desired.

**b)** For $n = 0$, we have

$$\Gamma(n+1) = \Gamma(1) = \int_0^\infty e^{-t}\, dt = 1 = 0! = n!,$$

and therefore a) implies that

$$\Gamma(1+1) = 1\Gamma(1) = 1 \cdot 0! = 1!,$$
$$\Gamma(2+1) = 2\Gamma(2) = 2 \cdot 1! = 2!,$$
$$\Gamma(3+1) = 3\Gamma(3) = 3 \cdot 2! = 3!,$$

and so forth. □

*Remark.* In Chapter 2, p.86ff, we used Stirling's formula

$$\lim_{n\to\infty} \frac{n! e^n}{n^n \sqrt{n}} = \sqrt{2\pi}$$

in order to make plausible the fact that binomial probabilities can be approximated by standard normal integrals whenever $n$ is sufficiently large. In the light of Theorem 4.5.2, we may write this formula in the following somewhat more general form, using a continuous variable $x$ instead of the discrete variable $n$:

$$\lim_{x\to\infty} \frac{\Gamma(x+1)e^x}{x^x \sqrt{x}} = \sqrt{2\pi}. \tag{4.27}$$

In order to derive this formula (in a semi-rigorous manner), we observe that

$$\Gamma(x+1) = \int_0^\infty t^x e^{-t}\, dt$$

and introduce the substitution $u = (t-x)/\sqrt{n}x$ with $du = dt/\sqrt{x}$. This yields

$$\Gamma(x+1) = \sqrt{x} \int_{-\sqrt{x}}^\infty (x + u\sqrt{x})^x e^{-(x+u\sqrt{x})}\, du$$

$$= \frac{x^x \sqrt{x}}{e^x} \int_{-\sqrt{x}}^\infty \left(1 + \frac{u}{\sqrt{x}}\right)^x e^{-u\sqrt{x}}\, du,$$

and as we now set

$$L := \lim_{x \to \infty} \left(1 + \frac{u}{\sqrt{x}}\right)^x e^{-u\sqrt{x}} = \lim_{y \to \infty} \left(\left(1 + \frac{u}{y}\right)^y e^{-u}\right)^y,$$

we may apply L'Hôpital's rule to conclude that

$$\ln(L) = \lim_{y \to \infty} \frac{y \ln(1 + u/y) - u}{1/y} = \lim_{y \to \infty} \frac{\ln(1 + u/y) - (u/y)/(1 + u/y)}{-1/y^2}$$

$$= -\lim_{y \to \infty} \frac{y \ln(1 + u/y) - u/(1 + u/y)}{1/y}$$

$$= -\ln(L) - u \lim_{y \to \infty} \frac{1 - 1/(1 + u/y)}{1/y} = -\ln(L) - u^2 \lim_{y \to \infty} \frac{1}{1 + u/y}$$

$$= -\ln(L) - u^2.$$

Hence

$$L = e^{\ln(L)} = e^{-u^2/2},$$

and therefore, for all sufficiently large values $x$ it is the case that

$$\Gamma(x + 1) \approx \frac{x^x \sqrt{x}}{e^x} \int_{-\sqrt{x}}^{\infty} e^{-u^2/2} \, du \approx \frac{x^x \sqrt{x}}{e^x} \int_{-\infty}^{\infty} e^{-u^2/2} \, du = \frac{x^x \sqrt{x} \sqrt{2\pi}}{e^x}.$$

Consequently, it is very much plausible to conclude that equation (4.27) is indeed valid.

**4.5.3 Definition.** A random variable $X : S \to \mathbb{R}$ is said to have a *gamma distribution* with parameters $\alpha, \beta > 0$ if its density is

$$f(x) = \begin{cases} \dfrac{x^{\alpha-1} e^{-x/\beta}}{\Gamma(\alpha)\beta^\alpha} & \text{if } x > 0, \\ 0 & \text{if } x \leq 0. \end{cases}$$

Furthermore, $X$ is said to have a *chi-squared distribution* with $\gamma$ degrees of freedom (for some positive integer $\gamma$), if $X$ is a gamma random variable with parameters $\alpha = \gamma/2$ and $\beta = 2$. In other words, the distribution of a chi-squared random variable is of the form

$$f(x) = \begin{cases} \dfrac{x^{\gamma/2-1} e^{-x/2}}{\Gamma(\gamma/2) 2^{\gamma/2}} & \text{if } x > 0, \\ 0 & \text{if } x \leq 0. \end{cases}$$

*Remark.* The fact that a gamma density satisfies the normalization condition $\int_{-\infty}^{\infty} f(x) \, dx = 1$ can be verified as follows:

$$\int_{-\infty}^{\infty} f(x) \, dx = \int_{0}^{\infty} \frac{x^{\alpha-1} e^{-x/\beta}}{\Gamma(\alpha)\beta^\alpha} \, dx = \int_{0}^{\infty} \frac{(\beta u)^{\alpha-1} e^{-u}}{\Gamma(\alpha)\beta^{\alpha-1}} \, du$$

$$= \int_{0}^{\infty} \frac{u^{\alpha-1} e^{-u}}{\Gamma(\alpha)} \, du = \frac{\Gamma(\alpha)}{\Gamma(\alpha)} = 1.$$

**4.5.4 Theorem.** *The moment-generating function of a gamma random variable $X$ with parameters $\alpha$ and $\beta$ is $m_X(t) = (1 - \beta t)^{-\alpha}$.*

*Proof.* If $t$ is less than $1/\beta$, then

$$
m_X(t) = \int_0^\infty \frac{e^{xt} x^{\alpha-1} e^{-x/\beta}}{\Gamma(\alpha)\beta^\alpha} \, dx = \int_{-\infty}^\infty \frac{x^{\alpha-1} e^{-(1/\beta - t)x}}{\Gamma(\alpha)\beta^\alpha} \, dx
$$

$$
= \int_0^\infty \frac{u^{\alpha-1} e^{-u}}{\Gamma(\alpha)\beta^\alpha (1/\beta - t)^{\alpha-1}} \frac{du}{1/\beta - t} = \int_0^\infty \frac{u^{\alpha-1} e^{-u}}{\Gamma(\alpha)(1 - \beta t)^\alpha} \, du
$$

$$
= \frac{\Gamma(\alpha)}{\Gamma(\alpha)(1 - \beta t)^\alpha} = (1 - \beta t)^{-\alpha},
$$

as desired.                                                                          □

**4.5.5 Corollary.** *The moment-generating function of a chi-squared random variable with $\gamma$ degrees of freedom is $m_X(t) = (1 - 2t)^{-\gamma/2}$.*

In order to determine the distribution of $S^2$, we will need the uniqueness property of the moment-generating function as stated in Theorem 4.5.6 below. As it turns out, this property is a direct consequence (in the continuous density case) of the uniqueness property of Laplace transforms, which is stated and proven in [K], Theorem 75.9, p.384.

**4.5.6 Theorem.** *Two random variables $X$ and $Y$ have the same distribution if there exists an $\varepsilon > 0$ such that $m_X(t) = m_Y(t)$ for all $t \in (-\varepsilon, \varepsilon)$.*

**4.5.7 Lemma.** *If $b$ is a real number, and $a$ is a positive real number, then*

$$
\int_{-\infty}^\infty \frac{e^{-ax^2 - bx}}{\sqrt{2\pi}} \, dx = \frac{e^{b^2/(4a)}}{\sqrt{2a}}.
$$

*Proof.* Setting $\sigma := 1/\sqrt{2a}$, we find that

$$
\int_{-\infty}^\infty \frac{e^{-ax^2 - bx}}{\sqrt{2\pi}} \, dx = \int_{-\infty}^\infty \frac{e^{-a((x+b/(2a))^2 - b^2/(4a^2))}}{\sqrt{2\pi}} \, dx
$$

$$
= e^{b^2/(4a)} \int_{-\infty}^\infty \frac{e^{-au^2}}{\sqrt{2\pi}} \, du = \sigma e^{b^2/(4a)} \int_{-\infty}^\infty \frac{e^{-u^2/(2\sigma^2)}}{\sqrt{2\pi}\sigma} \, du
$$

$$
= \frac{e^{b^2/(4a)}}{\sqrt{2a}},
$$

as desired.                                                                          □

The main result of this section, concerning the distribution of $(n-1)S_n^2$ (in the i.i.d. standard normal case) is stated in Theorem 4.5.8 below. As we will see, this theorem will allow us to compute $100q\%$ confidence intervals for the variance $\sigma^2$ of any normal random variable. However, as regards the proof of this theorem, we wish to caution that its details are somewhat cumbersome and may seem rather complicated to a beginning student of probability who has little experience with rigorous mathematical arguments. Moreover, the proof also involves a general proof pattern, known as mathematical induction, that may not be familiar to readers whose prior exposure to mathematics was limited to algebra and calculus. So unless a reader has the requisite background knowledge and truly desires to fully understand why Theorem 4.5.8 is valid, it may be best to move on to the subsequent discussion of how this theorem can be applied.

**4.5.8 Theorem.** *If $n > 1$ and if $X_1, \ldots, X_n$ are independent standard normal random variables, then the random variable*

$$(n-1)S_n^2 = \sum_{k=1}^{n} X_k^2 - n\overline{X}_n^2$$

*has a chi-squared distribution with $n-1$ degrees of freedom.*

*Proof.* According to Theorem 4.5.6 and Corollary 4.5.5, we need to show that the moment-generatingfunction of the random variable

$$X := \sum_{k=1}^{n} X_k^2 - n\overline{X}_n^2 = \sum_{k=1}^{n} X_k^2 - \frac{1}{n}\left(\sum_{k=1}^{n} X_k\right)^2$$

is

$$m_X(t) = (1-2t)^{-(n-1)/2}.$$

To do so, we will use induction to prove that for all $k \in \{0, \ldots, n-1\}$ it is the case that

$$m_X(t) = \int_{-\infty}^{\infty} \cdots \int_{-\infty}^{\infty} \frac{\sqrt{n}e^{f_{n,k}(t,x_1,\ldots,x_{n-k})}}{\sqrt{2\pi}^{n-k}h_{n,k}(t)} \, dx_{n-k} \ldots dx_1, \qquad (4.28)$$

where

$$f_{n,k}(t, x_1, \ldots, x_{n-k}) := -(1-2t)\sum_{i=1}^{n-k}\frac{x_i^2}{2} - \frac{t(1-2t)}{n-2t(n-k)}\left(\sum_{i=1}^{n-k} x_i\right)^2.$$

and

$$h_{n,k}(t) := \frac{\sqrt{n-2t(n-k)}\sqrt{1-2t}^{k-1}}{\sqrt{n}}.$$

For convenience, we also set

$$g_{n,k}(t, x_1, \ldots, x_{n-(k+1)}) := -(1 - 2t) \sum_{i=1}^{n-(k+1)} \frac{x_i^2}{2}$$

$$- \frac{t(1 - 2t)}{n - 2t(n - k)} \left( \sum_{i=1}^{n-(k+1)} x_i \right)^2.$$

Since $X_1, \ldots, X_n$ are independent standard normal random variables, we may apply Theorem 2.9.3 to infer that the joint density of these random variables is the product of the standard normal marginal densities. That is to say, the joint density is

$$\prod_{i=1}^{n} \frac{e^{-x_i^2/2}}{\sqrt{2\pi}} = \frac{e^{-\sum_{i=1}^{n} x_i^2/2}}{\sqrt{2\pi}^n}.$$

By implication, the moment-generating function $m_X(t) = E(e^{tX})$ is

$$m_X(t) = \int_{-\infty}^{\infty} \cdots \int_{-\infty}^{\infty} \frac{e^{t(\sum_{i=1}^{n} x_i^2 - (\sum_{i=1}^{n} x_i)^2/n) - (\sum_{i=1}^{n} x_i^2)/2}}{\sqrt{2\pi}^n} \, dx_n \ldots dx_1,$$

and the integral on the right is easily seen to be identical with the one in (4.28) for $k = 0$. In other words, the assertion in (4.28) is valid for $k = 0$. Assuming now that (4.28) is valid for some $k \in \{0, \ldots, k - 2\}$, we need to show—by way of induction—that it is valid for $k + 1$ as well. Taking a closer look at the exponent of the exponential function in the integrand in (4.28), we notice that

$$f_{n,k}(t, x_1, \ldots, x_{n-k}) = -(1 - 2t) \sum_{i=1}^{n-(k+1)} \frac{x_i^2}{2} - \frac{(1 - 2t)x_{n-k}^2}{2}$$

$$- \frac{t(1 - 2t)}{n - 2t(n - k)} \left( \left( \sum_{i=1}^{n-(k+1)} x_i \right)^2 + 2x_{n-k} \sum_{i=1}^{n-(k+1)} x_i + x_{n-k}^2 \right)$$

$$= g_{n,k}(t, x_1, \ldots, x_{n-(k+1)})$$

$$- \frac{t(1 - 2t)x_{n-k}^2}{n - 2t(n - k)} - \frac{(1 - 2t)x_{n-k}^2}{2} - \frac{2x_{n-k}t(1 - 2t)}{n - 2t(n - k)} \sum_{i=1}^{n-(k+1)} x_i$$

$$= g_{n,k}(t, x_1, \ldots, x_{n-(k+1)})$$

$$- \frac{(1 - 2t)x_{n-k}^2(n - 2t(n - (k + 1)))}{2(n - 2t(n - k))} - \frac{2x_{n-k}t(1 - 2t)}{n - 2t(n - k)} \sum_{i=1}^{n-(k+1)} x_i.$$

Consequently, we may use Lemma 4.5.7 with

$$a := \frac{(1 - 2t)(n - 2t(n - (k + 1)))}{2(n - 2t(n - k))}$$

and

$$b := \frac{2t(1-2t)}{n-2t(n-k)} \sum_{i=1}^{n-(k+1)} x_i,$$

to infer that

$$\int_{-\infty}^{\infty} \frac{e^{f_{n,k}(t,x_1,\ldots,x_{n-k})}}{\sqrt{2\pi} h_{n,k}(t)} \, dx_{n-k}$$

$$= \frac{e^{g_{n,k}(t,x_1,\ldots,x_{n-(k+1)})}}{h_{n,k}(t)} \int_{-\infty}^{\infty} \frac{e^{-ax_{n-k}^2 - bx_{n-k}}}{\sqrt{2\pi}} \, dx_{n-k}$$

$$= \frac{e^{g_{n,k}(t,x_1,\ldots,x_{n-(k+1)}) + b^2/(4a)}}{h_{n,k}(t)\sqrt{2a}} = \frac{e^{g_{n,k}(t,x_1,\ldots,x_{n-(k+1)}) + b^2/(4a)}}{h_{n,k+1}(t)}.$$

Furthermore, since

$$g_{n,k}(t, x_1, \ldots, x_{n-(k+1)}) + \frac{b^2}{4a}$$

$$= -(1-2t) \sum_{i=1}^{n-(k+1)} \frac{x_i^2}{2} - \frac{t(1-2t)}{n-2t(n-k)} \left( \sum_{i=1}^{n-(k+1)} x_i \right)^2$$

$$+ \frac{2t^2(1-2t)}{(n-2t(n-k))(n-2t(n-(k+1)))} \left( \sum_{i=1}^{n-(k+1)} x_i \right)^2$$

$$= -(1-2t) \sum_{i=1}^{n-(k+1)} \frac{x_i^2}{2}$$

$$- \frac{t(1-2t)(n-2t(n-(k+1))) - 2t)}{(n-2t(n-k))(n-2t(n-(k+1)))} \left( \sum_{i=1}^{n-(k+1)} x_i \right)^2$$

$$= f_{n,k+1}(t, x_1, \ldots, x_{n-(k+1)}),$$

it follows that

$$m_X(t) = \int_{-\infty}^{\infty} \cdots \int_{-\infty}^{\infty} \int_{-\infty}^{\infty} \frac{e^{f_{n,k}(t,x_1,\ldots,x_{n-k})}}{\sqrt{2\pi}^{\,n-(k+1)} \sqrt{2\pi} h_{n,k}(t)} \, dx_{n-k} \, dx_{n-(k+1)} \cdots dx_1$$

$$= \int_{-\infty}^{\infty} \cdots \int_{-\infty}^{\infty} \frac{e^{f_{n,k+1}(t,x_1,\ldots,x_{n-k})}}{\sqrt{2\pi}^{\,n-(k+1)} h_{n,k+1}(t)} \, dx_{n-(k+1)} \cdots dx_1.$$

Having thus established the validity of (4.28) for all $k \in \{0, \ldots, n-1\}$, we may set $k$ equal to $n-1$ to conclude that

$$m_X(t) = \int_{-\infty}^{\infty} \frac{e^{-(1-2t)x_1^2/2-t(1-2t)x_1^2/(n-2t)}}{\sqrt{2\pi}h_{n,n-1}(t)}\, dx_1$$

$$= \int_{-\infty}^{\infty} \frac{e^{-n(1-2t)x_1^2/(2(n-2t))}}{\sqrt{2\pi}h_{n,n-1}(t)}\, dx_1 = \frac{\sqrt{n-2t}}{h_{n,n-1}(t)\sqrt{n}\sqrt{1-2t}}$$

$$= (1-2t)^{-(n-1)/2},$$

as desired.                                                                 □

**4.5.9 Corollary.** *If $n > 1$ and if $X_1, \ldots, X_n$ are independent normal random variables with common mean $\mu$ and common standard deviation $\sigma$, then the random variable $(n-1)S^2/\sigma^2$ has a chi-squared distribution with $n-1$ degrees of freedom.*

*Proof.* Setting $Z_k := (X_k - \mu)/\sigma$, it follows that

$$\frac{(n-1)S^2}{\sigma^2} = \frac{1}{\sigma^2}\sum_{k=1}^{n}((X_k - \mu) - (\overline{X} - \mu))^2$$

$$= \frac{1}{\sigma^2}\sum_{k=1}^{n}((X_k - \mu)^2 - 2(X_k - \mu)(\overline{X} - \mu) + (\overline{X} - \mu)^2)$$

$$= \sum_{k=1}^{n} Z_k^2 - \frac{2(\overline{X} - \mu)}{\sigma}\sum_{k=1}^{n} Z_k + n\left(\frac{\overline{X} - \mu}{\sigma}\right)^2$$

$$= \sum_{k=1}^{n} Z_k^2 - \frac{2}{n}\left(\sum_{k=1}^{n} Z_k\right)^2 + \frac{1}{n}\left(\sum_{k=1}^{n} Z_k\right)^2$$

$$= \sum_{k=1}^{n} Z_k^2 - \frac{1}{n}\left(\sum_{k=1}^{n} Z_k\right)^2.$$

Thus, the statement of the corollary can be deduced by applying Theorem 4.5.8 to the independent *standard* normal random variables $Z_1, \ldots, Z_n$.       □

**4.5.10 Theorem.** *Assume that $X_1, \ldots, X_n$ (with $n > 1$) are i.i.d. normal random variables with standard deviation $\sigma$, and let $f$ be a chi-squared density with $n-1$ degrees of freedom. Then*

$$P\left(\frac{(n-1)S_n^2}{\beta} \le \sigma^2 \le \frac{(n-1)S_n^2}{\alpha}\right) = \int_{\alpha}^{\beta} f(x)\, dx.$$

*and*

$$P\left(\sigma^2 \le \frac{(n-1)S_n^2}{\alpha}\right) = \int_{\alpha}^{\infty} f(x)\, dx.$$

*for all $\alpha, \beta \in (0, \infty)$ with $\alpha < \beta$.*

*Proof.* Using Corollary 4.5.9, we find that

$$P\left(\frac{(n-1)S_n^2}{\beta} \le \sigma^2 \le \frac{(n-1)S_n^2}{\alpha}\right) = P\left(\frac{1}{\beta} \le \frac{\sigma^2}{(n-1)S_n^2} \le \frac{1}{\alpha}\right)$$

$$= P\left(\alpha \le \frac{(n-1)S_n^2}{\sigma^2} \le \beta\right) = \int_\alpha^\beta f(x)\, dx,$$

and the proof of the second equation above is completely analogous. $\qquad\square$

**4.5.11 Example.** Let us assume, as in Example 4.4.1, that $X : S \to \mathbb{R}$ is a normal random variable and that $x_1, \ldots, x_n$ is a corresponding random sample of output values from which we compute the sample variance $s_n^2$. Then, according to Theorem 4.5.10, the interval

$$\left[\frac{(n-1)s_n^2}{\beta}, \frac{(n-1)s_n^2}{\alpha}\right]$$

is a $100q\%$ confidence interval for $\sigma^2 = \mathrm{Var}(X)$ whenever $0 < \alpha < \beta$ and

$$q = \int_\alpha^\beta f(x)\, dx.$$

By implication, if we wish to find a 95% confidence interval for $\sigma^2$, then we may for instance solve for $\alpha$ the equation

$$0.025 = \int_0^\alpha f(x)\, dx$$

and for $\beta$ the equation

$$0.975 = \int_0^\beta f(x)\, dx$$

where $f$, as stated in Theorem 4.5.10, is a chi-squared density with $n - 1$ degrees of freedom, that is,

$$f(x) = \begin{cases} \dfrac{x^{(n-1)/2-1}e^{-x/2}}{\Gamma((n-1)/2)2^{(n-1)/2}} & \text{if } x > 0, \\ 0 & \text{if } x \le 0. \end{cases} \tag{4.29}$$

So if, for example, $n = 100$, then $\alpha$ is defined by the equation

$$0.025 = \int_0^\alpha \frac{x^{99/2-1}e^{-x/2}}{\Gamma(99/2)2^{99/2}}\, dx,$$

and $\beta$ by the equation

$$0.975 = \int_0^\beta \frac{x^{99/2-1}e^{-x/2}}{\Gamma(99/2)2^{99/2}}\, dx.$$

Using a computer, we find that

$$\alpha \approx 73.3611$$

and

$$\beta \approx 128.4220.$$

Moreover, if $n = 1000$, then we find in a completely analogous fashion that

$$\alpha \approx 913.3010 \tag{4.30}$$

and

$$\beta \approx 1088.4871.$$

Thus, the corresponding 95% confidence intervals are

$$\left[ \frac{99s_{100}^2}{128.4220}, \frac{99s_{100}^2}{73.3611} \right] \approx [0.7709s_{100}^2, 1.3495s_{100}^2]$$

and

$$\left[ \frac{999s_{1000}^2}{1088.4871}, \frac{999s_{100}^2}{913.3010} \right] \approx [0.9178s_{1000}^2, 1.0938s_{1000}^2],$$

respectively.

In Example 4.4.1 we showed how a $100q\%$ confidence interval for the mean $\mu$ of a normal random variable $X$ can be computed under the additional, somewhat artificial assumption that the variance of $X$ is known. In the example that follows we will drop this latter assumption and construct an interval $[a, b]$ in such a way that the probability for $\mu$ to be contained in this interval is, not equal to $q$, but *greater than or equal to* $q$. So the interval $[a, b]$ will be, so to speak, an upper estimate of an actual $100q\%$ confidence interval, and as such it will be, presumably, a bit larger than it would need to be ideally.

**4.5.12 Example.** Let $X : S \to \mathbb{R}$ be a normal random variable with mean $\mu$ and variance $\sigma^2$, and let $x_1, \ldots, x_n$ be a corresponding sample of output values. Given these assumptions, we wish to construct an interval $[a, b]$ such that the probability for $\mu$ to be contained in $[a, b]$ is greater than or equal to a given value $q \in (0, 1)$. To do so, we set

$$r := \frac{q + 1}{2}$$

and use (4.26) to infer that

$$[\bar{x}_n - \varepsilon, \bar{x}_n + \varepsilon]$$

is a $100r\%$ confidence interval for $\mu$ if the value $\varepsilon$ satisfies the equation

$$\frac{1 - r}{2} = \int_{-\infty}^{-\varepsilon\sqrt{n}/\sigma} \frac{e^{-x^2/2}}{\sqrt{2\pi}} \, dx.$$

Denoting by $y$ the solution of the equation

$$\frac{1-r}{2} = \int_{-\infty}^{-y} \frac{e^{-x^2/2}}{\sqrt{2\pi}}\, dx,$$

it follows that

$$y = \frac{\varepsilon\sqrt{n}}{\sigma},$$

and therefore, the confidence interval in question is

$$\left[\bar{x}_n - \frac{y\sigma}{\sqrt{n}}, \bar{x}_n + \frac{y\sigma}{\sqrt{n}}\right]. \tag{4.31}$$

Furthermore, if $f$ is a chi-squared density with $n-1$ degrees of freedom, then, according to Theorem 4.5.10, the interval

$$\left[0, \frac{(n-1)s_n^2}{\alpha}\right] \tag{4.32}$$

is a $100r\%$ confidence interval for $\sigma^2$ if $z$ satisfies the equation

$$r = \int_{\alpha}^{\infty} f(x)\, dx.$$

So what does this mean? Well, it means the following: if we generated samples of size $n$ repeatedly by means of the random process that $X$ describes, then we could expect that $100r\%$ of these samples produce a sample average $\bar{x}_n$ for which $\mu$ is contained in the interval (4.31) and that another $100r\%$ of these samples could be expected to produce a sample variance $s_n^2$ for which $\sigma^2$ is contained in the interval (4.32). So if we denote by $E$ the event that $\mu$ is contained in

$$\left[\bar{X}_n - \frac{y\sigma}{\sqrt{n}}, \bar{X}_n + \frac{y\sigma}{\sqrt{n}}\right].$$

and by $F$ the event that $\sigma^2$ is contained in

$$\left[0, \frac{(n-1)S_n^2}{\alpha}\right],$$

then

$$P(E \cap F) = P(E) + P(F) - P(E \cup F) \geq P(E) + P(F) - 1 = 2r - 1 = q.$$

Consequently, the joint probability for $\mu$ to be contained in the interval (4.31) *and* for $\sigma^2$ to be contained in the interval (4.32) is greater than or equal to $q$.

But if both of these conditions are satisfied simultaneously, then it also is the case that

$$\overline{x}_n - \frac{y\sqrt{n-1}s_n}{\sqrt{n}\sqrt{\alpha}} \le \overline{x}_n - \frac{y\sigma}{\sqrt{n}} \le \mu \le \overline{x}_n + \frac{y\sigma}{\sqrt{n}} \le \overline{x}_n + \frac{y\sqrt{n-1}s_n}{\sqrt{n}\sqrt{\alpha}},$$

and therefore, the chance for $\mu$ to be contained in the interval

$$[a,b] = \left[\overline{x}_n - \frac{y\sqrt{n-1}s_n}{\sqrt{n}\sqrt{\alpha}}, \overline{x}_n + \frac{y\sqrt{n-1}s_n}{\sqrt{n}\sqrt{\alpha}}\right]$$

is greater than or equal to $100q\%$, as desired. For instance, for $n = 1000$ and $q = 0.95$ we find that $r = 0.975$, and using a computer to approximately solve for $y$ and $\alpha$ the equations

$$\frac{1-r}{2} = \frac{1}{80} = \int_{-\infty}^{-y} \frac{e^{-x^2/2}}{\sqrt{2\pi}}\, dx$$

and

$$r = 0.975 = \int_{\alpha}^{\infty} f(x)\, dx$$

yields

$$y \approx 2.2414$$

and (as in (4.30))

$$\alpha \approx 913.3010,$$

respectively. Thus the chance for $\mu$ to be contained in the interval

$$\left[\overline{x}_{1000} - \frac{2.2414\sqrt{999}s_{1000}}{\sqrt{1000}\sqrt{913.3010}}, \overline{x}_{1000} + \frac{2.2414\sqrt{999}s_{1000}}{\sqrt{1000}\sqrt{913.3010}}\right] \tag{4.33}$$

$$\approx [\overline{x}_{1000} - 0.0741s_{1000}, \overline{x}_{1000} + 0.0741s_{1000}]$$

is greater than or equal to 95%.

*Remark.* The term $\Gamma((n-1)/2)$ that appears in (4.29) can be evaluated by means of Theorem 4.5.2 as follows: if $n = 2k+1$ is odd, then

$$\Gamma\left(\frac{n-1}{2}\right) = \Gamma(k) = (k-1)! = \left(\frac{n-3}{2}\right)!$$

and if $n = 2k$ is even, then

$$\Gamma\left(\frac{n-1}{2}\right) = \Gamma\left(k - \frac{1}{2}\right) = \Gamma\left(\frac{1}{2}\right)\prod_{i=1}^{k-1}\left(i - \frac{1}{2}\right) = \frac{2\Gamma(1/2)\prod_{i=1}^{k}(2i-1)}{(2k-1)2^k}$$

$$= \frac{\Gamma(1/2)(2k)!}{(2k-1)2^{k-1}\prod_{i=1}^{k}(2i)} = \frac{\Gamma(1/2)(2k)!}{(2k-1)2^{2k-1}k!}.$$

Furthermore, since

$$\Gamma\left(\frac{1}{2}\right) = \int_0^\infty x^{-1/2}e^{-x}\,dx = \int_0^\infty 2e^{-u^2}\,du = \int_{-\infty}^\infty e^{-u^2}\,du = \sqrt{\pi},$$

it follows in this latter case that

$$\Gamma\left(\frac{n-1}{2}\right) = \frac{\sqrt{\pi}(2k)!}{(2k-1)2^{2k-1}k!} = \frac{n!\sqrt{\pi}}{(n-1)2^{n-1}(n/2)!}. \qquad (4.34)$$

To conclude this section we wish to add the following fundamental fact concerning chi-squared distributions:

**4.5.13 Theorem.** *If $X_1, \ldots, X_n : S \to \mathbb{R}$ are independent standard normal random variables then $Y := X_1^2 + \cdots + X_n^2$ is a chi-squared random variable with $n$ degrees of freedom.*

*Proof.* According to Theorem 4.5.6, we only need to show that

$$m_Y(t) = (1-2t)^{-n/2}$$

for all $t$ that are sufficiently close to zero. Since $X_1, \ldots, X_n$ are assumed to be independent and standard normal, it follows that the joint density of these random variables is

$$f(x_1, \ldots, x_n) = \prod_{k=1}^n \frac{e^{-x_k^2/2}}{\sqrt{2\pi}} = \frac{e^{-(x_1^2+\cdots+x_n^2)/2}}{\sqrt{2\pi}^n}.$$

Hence

$$m_Y(t) = E(e^{tY}) = \int_{\mathbb{R}^n} e^{t(x_1^2+\cdots+x_n^2)} f(x_1, \ldots, x_n)\,dx_1 \ldots dx_n$$

$$= \prod_{k=1}^n \int_{-\infty}^\infty \frac{e^{-(1-2t)x_k^2/2}}{\sqrt{2\pi}}\,dx_k = \prod_{k=1}^n \frac{1}{\sqrt{1-2t}} = (1-2t)^{-n/2}$$

for all $t < 1/2$, as desired. $\qquad \square$

*Remark.* Since

$$\frac{1}{n}\sum_{k=2}^n \sum_{i=1}^{k-1}(X_k - X_i)^2 = \frac{1}{n}\left(\sum_{k=2}^n \sum_{i=1}^{k-1} X_k^2 - \sum_{k=2}^n \sum_{i=1}^{k-1} 2X_k X_i + \sum_{k=2}^n \sum_{i=1}^{k-1} X_i^2\right)$$

$$= \frac{1}{n}\left(\sum_{k=2}^n (k-1)X_k^2 - \sum_{k=2}^n \sum_{i=1}^{k-1} 2X_k X_i + \sum_{i=1}^{n-1} \sum_{k=i+1}^n X_i^2\right)$$

$$= \frac{1}{n}\left(\sum_{k=2}^n (k-1)X_k^2 - \sum_{k=2}^n \sum_{i=1}^{k-1} 2X_k X_i + \sum_{i=1}^{n-1} (n-i)X_i^2\right)$$

$$= \frac{1}{n} \left( \sum_{k=1}^{n}(k-1)X_k^2 - \sum_{k=2}^{n}\sum_{i=1}^{k-1}2X_kX_i + \sum_{k=1}^{n}(n-k)X_k^2 \right)$$

$$= \frac{1}{n} \left( (n-1)\sum_{k=1}^{n}X_k^2 - \sum_{k=2}^{n}\sum_{i=1}^{k-1}2X_kX_i \right)$$

$$= \frac{1}{n} \left( n\sum_{k=1}^{n}X_k^2 - \left( \sum_{k=1}^{n}X_k^2 + \sum_{k=2}^{n}\sum_{i=1}^{k-1}2X_kX_i \right) \right)$$

$$= \sum_{k=1}^{n}X_k^2 - \frac{1}{n}\left( \sum_{k=1}^{n}\sum_{i=1}^{n}X_kX_i \right)$$

$$= \sum_{k=1}^{n}X_k^2 - \frac{1}{n}\left( \sum_{k=1}^{n}X_k \right)^2$$

$$= \sum_{k=1}^{n}X_k^2 - n\overline{X}_n^2 = (n-1)S_n^2,$$

it may appear that the fact that $(n-1)S_n^2$ has a chi-squared distribution can be deduced from Theorem 4.5.13 merely by observing that $(X_k - X_i)/\sqrt{2}$ is standard normal whenever $i \neq k$. But here we would be mistaken because the random variables $X_k - X_i$ with $2 \leq k \leq n$ and $1 \leq i \leq k-1$ are not independent.

## Exercises

**4.5.14.** Assume that the sample variance of a normal random process turned out to be $s_{400}^2 = 4.14$ after $n = 400$ trials.

   a) Find a 95% confidence interval for $\sigma^2$.

   b) Find a 97% confidence interval for $\sigma^2$.

**4.5.15.** Assume that the sample mean and the sample variance of a normal random process turned out to be $\overline{x}_{400} = 18.32$ and $s_{400}^2 = 4.14$ after $n = 400$ trials.

   a) Find a 95% confidence interval for $\mu$.

   b) Find a 97% confidence interval for $\mu$.

**4.5.16.** The random variables $X_1, \ldots, X_{10000}$, that may be thought to be associated with the 10000 outcomes $x_1, \ldots, x_{10000}$ in the table in Exercise 4.3.1, are approximately normal because they are binomial with parameters $n = 200$ and $p \approx 0.4$. Use this observation to find approximate 95% confidence intervals for $\sigma^2 = \text{Var}(X_k)$ and $\mu = E(X_k)$.

**4.5.17.** Assume that the density of a random variable $X$ is the gamma density

$$f(x) = \begin{cases} x^n e^{-x}/n! & \text{if } x > 0 \\ 0 & \text{if } x \leq 0 \end{cases}$$

(with parameters $\alpha = n + 1$ and $\beta = 1$). Use Theorem 4.5.2 to find $E(X)$ and $\text{Var}(X)$.

## 4.6   Estimating $\mu$ if $\sigma$ Is Not Known

In the previous section we constructed an upper estimate for a confidence interval for $\mu$ by combining an estimate for $\mu$ with an estimate for $\sigma^2$. In order to improve this method and to find an exact $100q\%$ confidence interval for $\mu$ in the case where $\sigma$ is not known, we need to introduce the following definition:

**4.6.1 Definition.** A random variable $X : S \to \mathbb{R}$ is said to be a *T random variable* with $\gamma$ degrees of freedom (for some $\gamma \in \mathbb{N}$) if its density (defined on $\mathbb{R}$) is

$$f(x) = \frac{\Gamma((\gamma+1)/2)}{\Gamma(\gamma/2)\sqrt{\gamma\pi}} \left(1 + \frac{x^2}{\gamma}\right)^{-(\gamma+1)/2}.$$

*Remark.* The fact that this function $f$ satisfies the normalization condition $\int_{-\infty}^{\infty} f(x)\,dx = 1$ can be verified as follows: using integration by parts and assuming that $\alpha > 3/2$, we find that

$$\int_{-\infty}^{\infty} \left(1 + \frac{x^2}{\gamma}\right)^{-\alpha+1} dx =$$

$$= x\left(1 + \frac{x^2}{\gamma}\right)^{-\alpha+1} \Bigg|_{-\infty}^{\infty} - (-\alpha+1)\int_{-\infty}^{\infty} \frac{2x^2}{\gamma}\left(1 + \frac{x^2}{\gamma}\right)^{-\alpha} dx$$

$$= 2(\alpha-1)\left(\int_{-\infty}^{\infty}\left(1 + \frac{x^2}{\gamma}\right)^{-\alpha+1} dx - \int_{-\infty}^{\infty}\left(1 + \frac{x^2}{\gamma}\right)^{-\alpha} dx\right),$$

and, by implication,

$$\int_{-\infty}^{\infty}\left(1 + \frac{x^2}{\gamma}\right)^{-\alpha} dx = \frac{2(\alpha-1)-1}{2(\alpha-1)}\int_{-\infty}^{\infty}\left(1 + \frac{x^2}{\gamma}\right)^{-(\alpha-1)} dx.$$

Hence the function

$$I(a) := \int_{-\infty}^{\infty}\left(1 + \frac{x^2}{\gamma}\right)^{-(a+1)/2} dx,$$

satisfies the equation

$$I(a) = \frac{a-2}{a-1} \cdot I(a-2)$$

for all integers $a > 2$. The same equation is satisfied as well for the function

$$J(a) := \frac{\Gamma(a/2)}{\Gamma((a+1)/2)}$$

because

$$J(a) = \frac{(a/2-1)\Gamma(a/2-1)}{((a+1)/2-1)\Gamma((a+1)/2-1)} = \frac{a-2}{a-1} \cdot J(a-2).$$

Consequently, for all $a \in \mathbb{N}$ we have

$$\frac{I(a)}{J(a)} = \begin{cases} I(1)/J(1) & \text{if } a \text{ is odd,} \\ I(2)/J(2) & \text{if } a \text{ is even.} \end{cases}$$

Using elementary techniques of integration, it is easy to see that $I(2) = 2\sqrt{\gamma}$, $I(1) = \pi\sqrt{\gamma}$, and $\Gamma(3/2) = \Gamma(1/2)/2 = \sqrt{\pi}/2$. Thus, $J(1) = \sqrt{\pi}$, $J(2) = 2/\sqrt{\pi}$, and

$$\frac{I(1)}{J(1)} = \frac{I(2)}{J(2)} = \sqrt{\gamma\pi}.$$

This shows that

$$\frac{I(a)}{J(a)} = \sqrt{\gamma\pi},$$

for all $a \in \mathbb{N}$. Hence

$$\int_{-\infty}^{\infty} f(x)\, dx = \int_{-\infty}^{\infty} \frac{\Gamma((\gamma+1)/2)}{\Gamma(\gamma/2)\sqrt{\gamma\pi}} \left(1 + \frac{x^2}{\gamma}\right)^{-(\gamma+1)/2} dx$$

$$= \frac{1}{\sqrt{\gamma\pi}} \left(\left.\frac{I(a)}{J(a)}\right|_{a=\gamma}\right) = 1,$$

as desired.

In order not to get bogged down in another cumbersome proof, similar to the proof of Theorem 4.5.8, we will not attempt to derive the statement of Theorem 4.6.2 below—the main result in this concluding section—and focus our attention instead on how it can be applied.

**4.6.2 Theorem.** *If $n > 1$ and if $X_1, \ldots, X_n : S \to \mathbb{R}$ are i.i.d. normal random variables, then*

$$\frac{\overline{X}_n - \mu}{S_n/\sqrt{n}}$$

*is a $T$ random variable with $n - 1$ degrees of freedom.*

**4.6.3 Theorem.** *Assume that $X_1, \ldots, X_n : S \to \mathbb{R}$ are i.i.d. normal random variables (for some $n > 1$) and let $f$ be a $T$ density with $n - 1$ degrees of freedom. Then*

$$P\left(|\overline{X}_n - \mu| \leq \frac{\varepsilon S_n}{\sqrt{n}}\right) = \int_{-\varepsilon}^{\varepsilon} f(x)\,dx.$$

*for all $\varepsilon > 0$.*

*Proof.* Using Theorem 4.6.2, it follows that

$$\int_{-\varepsilon}^{\varepsilon} f(x)\,dx = P\left(-\varepsilon \leq \frac{\overline{X}_n - \mu}{S_n/\sqrt{n}} \leq \varepsilon\right) = P\left(|\overline{X}_n - \mu| \leq \frac{\varepsilon S_n}{\sqrt{n}}\right),$$

as desired.                                                                            □

Since the inequality

$$|\overline{X}_n - \mu| \leq \frac{\varepsilon S_n}{\sqrt{n}}$$

is equivalent to the dual inequality

$$\overline{X}_n - \frac{\varepsilon S_n}{\sqrt{n}} \leq \mu \leq \overline{X}_n + \frac{\varepsilon S_n}{\sqrt{n}},$$

we may apply Theorem 4.6.3 to infer that for any random sample $x_1, \ldots, x_n$ that consists of potential output values of a normal random variable $X : S \to \mathbb{R}$ the interval

$$\left[\overline{x}_n - \frac{\varepsilon s_n}{\sqrt{n}}, \overline{x}_n + \frac{\varepsilon s_n}{\sqrt{n}}\right]$$

is a $100q\%$ confidence interval for the mean $\mu = E(X)$ whenever $q$ is equal to the $T$ density integral given above, that is,

$$q = \int_{-\varepsilon}^{\varepsilon} f(x)\,dx.$$

**4.6.4 Example.** Given the sample size $n = 1000$, we wish to find a 95% confidence interval for the mean $\mu$ of a normal random variable $X : S \to \mathbb{R}$. To do so, we need to solve for $\varepsilon$ the equation

$$q = 0.95 = \int_{-\varepsilon}^{\varepsilon} f(x)\,dx = \int_{-\varepsilon}^{\varepsilon} \frac{\Gamma(n/2)}{\Gamma((n-1)/2)\sqrt{(n-1)\pi}} \left(1 + \frac{x^2}{n-1}\right)^{-n/2} dx$$

$$= 2\int_{0}^{\varepsilon} \frac{\Gamma(500)}{\Gamma(999/2)\sqrt{999\pi}} \left(1 + \frac{x^2}{999}\right)^{-500} dx.$$

Using a computer, we find the (rounded) approximate solution

$$\varepsilon \approx 1.9623.$$

Thus the desired confidence interval is

$$\left[\overline{x}_n - \frac{1.9623 s_n}{\sqrt{1000}}, \overline{x}_n + \frac{1.9623 s_n}{\sqrt{1000}}\right] \approx \left[\overline{x}_n - 0.0621 s_n, \overline{x}_n + 0.0621 s_n\right]$$

which is slightly narrower—as it ought to be—than the interval (4.33) in Example 4.5.12.

*Remark.* If $X_1, \ldots, X_n : S \to \mathbb{R}$ are i.i.d. normal random variables with (common) mean $\mu$ and (common) standard deviation $\sigma$, then, $\overline{X}_n$ is normal with mean $\mu$ and standard deviation $\sigma/\sqrt{n}$ (by Proposition 3.2.6). Consequently, according to Corollary 2.6.7, the random variable

$$\frac{\overline{X}_n - \mu}{\sigma/\sqrt{n}}$$

is standard normal. Since $S_n^2$ is an estimator for $\sigma^2$ that is ever more likely to be ever more accurate as $n$ increases, we ought to expect that the $T$ density of

$$\frac{\overline{X}_n - \mu}{S_n/\sqrt{n}}$$

approaches the standard normal density $z(x)$ as $n$ tends to infinity. That is to say, it ought to be the case that

$$\lim_{n \to \infty} \frac{\Gamma(n/2)}{\Gamma((n-1)/2)\sqrt{(n-1)\pi}}\left(1 + \frac{x^2}{n-1}\right)^{-n/2} = \frac{e^{-x^2/2}}{\sqrt{2\pi}}.$$

for all $x \in \mathbb{R}$. In order to see why this indeed is true, we apply (4.27) to infer that

$$\lim_{n \to \infty} \frac{\Gamma(n/2)}{\Gamma((n-1)/2)\sqrt{(n-1)\pi}} = \lim_{n \to \infty} \frac{\dfrac{\sqrt{2\pi}(n/2)^{n/2}}{e^{n/2}\sqrt{n/2}}}{\dfrac{\sqrt{2\pi}((n-1)/2)^{(n-1)/2}\sqrt{(n-1)\pi}}{e^{(n-1)/2}\sqrt{(n-1)/2}}}$$

$$= \lim_{n \to \infty} \frac{(n/2)^{n/2}e^{-1/2}}{\sqrt{n\pi}((n-1)/2)^{(n-1)/2}}$$

$$= \lim_{n \to \infty} \frac{n^{n/2}e^{-1/2}\sqrt{(n-1)/2}}{\sqrt{n\pi}(n-1)^{n/2}}$$

$$= \lim_{n \to \infty} \frac{e^{-1/2}}{\sqrt{2\pi}(1-1/n)^{n/2}}$$

$$= \frac{1}{\sqrt{2\pi}} \quad \text{(by (2.6))}.$$

Hence

$$\lim_{n\to\infty} \frac{\Gamma(n/2)}{\Gamma((n-1)/2)\sqrt{(n-1)\pi}} \left(1+\frac{x^2}{n-1}\right)^{-n/2} = \lim_{n\to\infty} \frac{\left(1+\frac{x^2}{n-1}\right)^{-n/2}}{\sqrt{2\pi}}$$

$$= \frac{1}{\sqrt{2\pi}} \lim_{n\to\infty} \left(1+\frac{x^2}{n}\right)^{-n/2} = \frac{e^{-x^2/2}}{\sqrt{2\pi}} \quad \text{(by (2.6))},$$

as conjectured.

# Exercises

**4.6.5.** Find a 95% confidence interval for the mean $\mu$ of a normal random variable $X$ under the assumption that $\overline{x}_{400} = 3.45$ and $s^2_{400} = 1.23$.

**4.6.6.** Find a 99% confidence interval for the mean $\mu$ of a normal random variable $X$ under the assumption that $\overline{x}_{900} = 13.15$ and $s^2_{900} = 2.46$.

**4.6.7.** Assume that the values $x_1 = 5.1$, $x_2 = 6.1$, $x_3 = 6.3$, $x_4 = 5.9$, $x_5 = 7.5$, $x_6 = 6.7$, $x_7 = 6.9$, $x_8 = 4.6$, $x_9 = 7.0$, $x_{10} = 6.2$, $x_{11} = 5.5$, $x_{12} = 5.1$, $x_{13} = 6.6$, $x_{14} = 5.8$, $x_{15} = 5.2$, $x_{16} = 4.9$, $x_{17} = 7.3$, $x_{18} = 6.4$, $x_{19} = 5.6$, $x_{20} = 5.8$ are a random sample that consists of specific output values of a random process that is described by a normal random variable $X$.

a) Given the values above, what is the presumably best estimate for $\mu = E(X)$?

b) What is the presumably best estimate for the variance $\sigma^2 = \text{Var}(X)$?

c) Find a 95% confidence interval for $\sigma^2$.

d) Use a $T$ density to find a 97% confidence interval for $\mu$.

# Selected Solutions

## Chapter 1

**1.1.7** $\{1, 2, 3, 4\}$ and $\{1, 2\}$ are subsets.

**1.1.8** $\{1, 2\} \cup \{2, 5, 6\} = \{1, 2, 5, 6\}$, $\{1, 2, 3, 4\} \cap \{2, 4, 7\} = \{2, 4\}$, $\{1, 2, 3, 4\} \smallsetminus \{2, 4, 7\} = \{1, 3\}$.

**1.1.9** $-1$ and $1/2$.

**1.1.10** $[1, 2) \cup [2, 3] = [1, 3]$, $[1, 2] \smallsetminus \{2\} = [1, 2)$, $[1, 2] \smallsetminus \{1, 2\} = (1, 2)$, $[0, 3] \smallsetminus [1, 3] = [0, 1)$, $\mathbb{R} \smallsetminus (-\infty, 0) = [0, \infty)$, $[1, 2] \cap (3/2, 3) = (3/2, 2]$, $[1, 2] \cap [2, 3] = [2, 2]$, and $(a, b) \cup (b, a) = ((a + b - |a - b|)/2, (a + b + |a - b|)/2)$.

**1.1.12 a)** $D(f) = \mathbb{R} \smallsetminus \{2\}$, **c)** $f(f(1)) = f(-1) = -2/3$.

**1.1.13** $D(f) = (-\infty, -2] \cup [2, \infty) = \mathbb{R} \smallsetminus (-2, 2)$

**1.1.14** The equation $y^2 = x^2$ is equivalent to the equation $y = \pm x$. The latter equation does not define $y$ as a function of $x$ on $\mathbb{R}$ because it assigns every value $x \neq 0$ two distinct values of $y$.

**1.1.15 c)** $(0, \pi/2)$, **d)** $\{(x, y) \in \mathbb{R}^2 \mid x^2 + y^2 < 1\}$.

**1.1.16 a)** $\{1\}$, **c)** $[-1, 1]$.

**1.1.17 a)** $A = B$, **c)** $B \subset A$ because the empty set is a subset of every set.

**1.1.18 a)** $D(f) = [1, 4]$, $R(f) = [0, 3]$.

**1.2.11** $6^3 = 216$

**1.2.12** $\{3, 4, \ldots, 18\}$

**1.2.13** $\{(k, i) \mid k \in \{1, \ldots, 10\} \wedge i \in \{k, \ldots, 10\}\}$

**1.2.14** $\{2, 3, \ldots, 20\}$

**1.3.24**

$$
\begin{aligned}
\binom{n}{k-1} + \binom{n}{k} &= \frac{n!}{(k-1)!(n-k+1)!} + \frac{n!}{k!(n-k)!} \\
&= \frac{n!k}{k!(n-k+1)!} + \frac{n!(n-k+1)}{k!(n-k+1)!} \\
&= \frac{n!(n+1)}{k!(n+1-k)!} = \frac{(n+1)!}{k!(n+1-k)!} = \binom{n+1}{k}.
\end{aligned}
$$

**1.3.26** $40/500 = 2/25$

**1.3.28 a)** $2^5 = 32$, **b)** $2^n$, **c)** $2^n$

**1.3.29**

$$\frac{\binom{26}{6}}{\binom{30}{6}} = \frac{26!6!24!}{6!20!30!} = \frac{24 \cdot 23 \cdot 22 \cdot 21}{30 \cdot 29 \cdot 28 \cdot 27} \approx 0.388$$

**1.3.30 a)** $3^9$ **b)** $\frac{1}{3^9} \cdot \frac{9!}{3!2!4!} \approx 0.064$

**1.3.31** $8^2 \binom{7}{2} = 64 \cdot 21 = 1344$

**1.3.32** If the balls are not returned, then the probability is

$$\frac{5}{8} \cdot \frac{3}{7} + \frac{3}{8} \cdot \frac{5}{7} = \frac{15}{28}$$

and otherwise it is

$$\frac{5}{8} \cdot \frac{3}{8} + \frac{3}{8} \cdot \frac{5}{8} = \frac{15}{32}.$$

**1.3.37** Denoting the probability by $p(a)$, we have

$$p(a) = \begin{cases} (1 - 2a)^2 & \text{if } a \in [0, 1/2], \\ 0 & \text{if } a \in [1/2, 1). \end{cases}$$

So the graph of $p$ is an upward parabola on the interval $[0, 1/2]$ with vertex at $(1/2, 0)$ and it coincides with the $a$-axis on the interval $[1/2, 1]$. Furthermore, $p(a) = 1/2$ for $a = (\sqrt{2} - 1)/(2\sqrt{2})$.

**1.3.38** The probability in question is the relative size of the area above the graph $x = 1/(2\sin(\theta))$ in the sample space $S = [0, \pi] \times [0, 1]$. Using a computer, we find that

$$P(E) = \frac{1}{\pi}\left(\frac{2\pi}{3} - \int_{\pi/6}^{5\pi/6} \frac{1}{2\sin(\theta)}\, d\theta\right) \approx 0.247.$$

**1.3.39** Setting $S = [0, 2] \times [0, 2]$ and using the condition $st \leq 1/2$, we find that

$$P(E) = \frac{1}{4}\left(\frac{2}{4} + \int_{1/4}^{2} \frac{1}{2s}\, ds\right) = \frac{1 + \ln(2) - \ln(1/4)}{8} = \frac{1 + 3\ln(2)}{8}.$$

**1.3.40** Since $r = x/(2\pi)$, it follows that $\pi(x/(2\pi))^2 < 9/\pi$ or, equivalently, $x < 6$. Hence $P(E) = 6/7$.

**1.4.4 c)** $P(F) = P(F \setminus E) + P(E)$, and since $P(F \setminus E) \geq 0$, it follows that $P(F) \geq P(E)$.

**d)** Since $E \subset S$, c) implies that $1 = P(S) \geq P(E)$.

**1.4.6** $P(E \cap F) = P(E) + P(F) - P(E \cap F) = 0.6 + 0.7 - (1 - 0.1) = 0.4$.

**1.4.7** $P(E \cup F \cup G) =$
$P(E) + P(F) + P(G) - P(E \cap F) - P(E \cap G) - P(F \cap G) + P(E \cap F \cap G)$

**1.4.9** Since $P((E \cap F) \cup (E \cap G)) = P(F \cup G)$, we may infer that $P((E \cup F \cup G) \setminus E) = 0$, and therefore $P(E) = P(E \cup F \cup G) = 3/4$.

**1.4.10** Since $0.7 = P(E \cup F) = P(E) + P(F) - P(E \cap F) = P(E) + P(F) - 0.5$, it follows that $P(F) = 1.2 - P(E)$, and therefore,
$P(E)P(F) = P(E)(1.2 - P(E)) \geq 0.5(1.2 - 0.5) = 0.7(1.2 - 0.7) = 0.35$ because $0.7 \geq P(E \cup F) \geq P(E) \geq P(E \cap F) = 0.5$.

**1.4.11** Setting $a := P(E \setminus F)$, $b := P(E \cap F)$, and $c := P(F \setminus E)$, it follows that $P(E)P(F) = (a+b)(b+c) = ab + ac + b^2 + bc \geq ab + b^2 + bc = b(a+b+c) = P(E \cap F)P(E \cup F)$.

**1.4.12** Using the fact that $P(E \cup F) = P(E) + P(F) - P(E \cap F)$, it follows that

$$\sqrt{P(E)P(F)} \leq \frac{P(E \cap F) + P(E \cup F)}{2}$$
$$\Leftrightarrow \sqrt{P(E)P(F)} \leq \frac{P(E) + P(F)}{2}$$
$$\Leftrightarrow 4P(E)P(F) \leq P(E)^2 + 2P(E)P(F) + P(F)^2$$
$$\Leftrightarrow 0 \leq P(E)^2 - 2P(E)P(F) + P(F)^2$$
$$\Leftrightarrow 0 \leq (P(E) - P(F))^2,$$

and the last of these inequalities is evidently valid.

**1.5.6** $P(E \cap F) = P(E) + P(F) - P(E \cup F) = 0.5$, and therefore, $P(E|F) = 0.5/0.6 = 5/6$ and $P(F|E) = 0.5/0.7 = 5/7$. Since $P(E|F) = 5/6 \neq 0.7 = P(E)$, it follows that $E$ and $F$ are not independent.

**1.5.8** $P(E \cap F) = P(E|F)P(F) = 0.5 \cdot 0.4 = 0.2$, and therefore, $P(E \cup F) = P(E) + P(F) - P(E \cap F) = 0.7$.

**1.5.9** If both $P(E)$ and $P(F)$ are greater than zero, the $P(E)P(F) \neq 0 = P(E \cap F)$, and therefore, $E$ and $F$ are not independent.

**1.5.10** If $P(E) = 0$, then $0 \leq P(E \cap F) \leq P(E) = 0$ for all events $F$. Hence $P(E \cap F) = 0 = P(E)P(F)$, as desired.

**1.5.14** $P(E) = P(\{(1,9), (2,8), (3,7), (4,6), (5,5)\}) = (1/10 + 1/9 + 1/8 + 1/7 + 1/6)/10 \approx 0.065$,
$P(F) = P(\{(1,3), (2,4), (3,5), (4,6), (5,7), (6,8), (7,9), (8,10)\}) = (1/10 + 1/9 + 1/8 + 1/7 + 1/6 + 1/5 + 1/4 + 1/3)/10 \approx 0.143$,

$P(E \cap F) = P(\{(4,6)\}) = (1/7)/10 \approx 0.014$, and therefore
$P(E|F) \approx 0.014/0.143 \approx 0.1 \neq 0.065 \approx P(E)$. Thus $E$ and $F$ are not independent.

# Chapter 2

**2.1.8** time and space

**2.1.9** a) and b)

**2.2.5 a)** $S = \{(k,i) \mid 1 \leq k \leq n \wedge 1 \leq i \leq k\}$
**b)** $f_X(i) = \sum_{j=i}^{n} 1/(jn)$ if $i \in \{1, \ldots, n\}$ and $f_X(i) = 0$ otherwise.
**c)** $f_Y(k) = 1/n$ for $k \in \{1, \ldots, n\}$ and $f_Y(k) = 0$ otherwise.
**d)** $f_Z(m) = \sum_{j=m+1}^{n} 1/(jn)$ if $m \in \{0, \ldots, n-1\}$ and $f_Z(m) = 0$ otherwise.
**2.2.6 a)** $R(X) = \mathbb{N}$
**b)** $f(k) = (2/3)^{k-1}/3$
**c)** $F(k) = \sum_{i=1}^{k} (2/3)^{i-1}/3 = \sum_{i=0}^{k-1} (2/3)^i/3 = 1 - (2/3)^k$ if $k \geq 1$ and $F(k) = 0$ if $k < 1$.

**2.2.7** $F(k) = 0$ if $k < 0$, $F(k) = 1$ if $k \geq 9$, and $F(k) = (k+1)/10$ if $0 \leq k \leq 8$.

**2.3.16** $E(X) = \sum_{i=1}^{n} i f_X(i) = \sum_{i=1}^{n} \sum_{j=i}^{n} i/(jn) = \sum_{j=1}^{n} \sum_{i=1}^{j} i/(jn) = \sum_{j=1}^{n} (j+1)/(2n) = (n+3)/4$

**2.3.17** $1/(1-x) = \sum_{k=1}^{\infty} x^k$ and therefore $1/(1-x)^2 = \sum_{k=1}^{\infty} k x^{k-1}$. Hence $E(X) = \sum_{k=1}^{\infty} k(2/3)^{k-1}/3 = (1/3)/(1-2/3)^2 = 3$.

**2.3.20 a)** Since

$$1 = C \sum_{k=1}^{\infty} \frac{1}{k(k+1)(k+2)} = C \sum_{k=1}^{\infty} \left( \frac{1}{2k} - \frac{1}{k+1} + \frac{1}{2(k+2)} \right) = \frac{C}{4},$$

it follows that $C = 4$.
**b)** $E(X) = 4 \sum_{k=1}^{\infty} k/(k(k+1)(k+2)) = 4 \sum_{k=1}^{\infty} 1/((k+1)(k+2)) = 2$
**c)** $E(X^2) = 4 \sum_{k=1}^{\infty} k^2/(k(k+1)(k+2)) = 4 \sum_{k=1}^{\infty} k/((k+1)(k+2)) = \infty$
**2.3.21** Denoting by $g$ and $h$ the densities of $2X$ and $X^2$ respectively, it follows that for any positive integer $k$ we have

$$g(k) = \begin{cases} f(k/2) & \text{whenever } k \text{ is even,} \\ 0 & \text{otherwise} \end{cases}$$

and

$$h(k) = \begin{cases} f\left(\sqrt{k}\right) & \text{whenever } k \text{ is a square,} \\ 0 & \text{otherwise.} \end{cases}$$

Thus, the corresponding expected values are $E(2X) = \sum_{k=1}^{\infty} k g(k) = \sum_{k=1}^{\infty} 2k f(2k/2) = 2 \sum_{k=1}^{\infty} k f(k) = 2E(X)$ and $E(X^2) = \sum_{k=1}^{\infty} k h(k) = \sum_{k=1}^{\infty} k^2 f\left(\sqrt{k^2}\right) = \sum_{k=1}^{\infty} k^2 f(k)$.

**2.3.22 a)** $E(2X) = \sum_{x \in R(X)} 2x f(x) = 2 \sum_{x \in R(X)} x f(x) = 2E(X)$

**b)** The equation $E(X^2) = E(X)^2$ is in general not satisfied. For example, if $f(1) = 1/2 = f(-1)$, then $E(X)^2 = (f(1) - f(-1))^2 = 0$ but $E(X^2) = f(1) + f(-1) = 1$. Furthermore, if it were satisfied, the variance would always be zero.

**2.3.23 a)** $1 = C \sum_{k=1}^{\infty} k/3^k = (C/3) \sum_{k=1}^{\infty} k/3^{k-1} = C/3/(1 - (1/3))^2 = 3C/4$, and therefore, $C = 4/3$.

**b)**

$$m_X(t) = \frac{4}{3} \sum_{k=1}^{\infty} \frac{e^{tk} k}{3^k} = \frac{4e^t}{9} \sum_{k=1}^{\infty} k \left(\frac{e^t}{3}\right)^{k-1} = \frac{4e^t}{9(1 - e^t/3)^2} = \frac{4e^t}{(3 - e^t)^2}$$

**c)**

$$m_X'(t) = \frac{4e^t(3 - e^t)^2 + 8e^{2t}(3 - e^t)}{(3 - e^t)^4} = \frac{12e^t + 4e^{2t}}{(3 - e^t)^3}$$

and therefore, $E(X) = m_X'(0) = 16/8 = 2$. $Var(X)$ is found similarly by computing $m_X''(0) = E(X^2)$.

**2.3.24 a)** $P(X = 2) = 1 - P(X = 0) - P(X = 1) = (1 - p)^2$

**b)** $m_X(t) = p^2 + 2p(1 - p)e^t + (1 - p)^2 e^{2t} = (p + e^t(1 - p))^2$

**c)** $m_X'(t) = 2e^t(1 - p)(p + e^t(1 - p)) = 2e^t(1 - p)p + 2e^{2t}(1 - p)^2$ and $m_X''(t) = 2e^t(1 - p)p + 4e^{2t}(1 - p)^2$, and therefore, $E(X) = m_X'(0) = 2(1 - p)$ and $Var(X) = 2(1 - p)p + 4(1 - p)^2 - 4(1 - p)^2 = 2p(1 - p)$

**2.3.26** Given that $Var(X) = 12 = np(1 - p)$, we may infer that $12/n$ must be less than or equal to the maximum of the function $f(p) = p(1 - p)$ on the interval $[0, 1]$. Since the value of this maximum is $f(1/2) = 1/4$, it follows that $12/n \leq 1/4$, and therefore, the smallest possible value of $n$ is 48.

**2.3.27** The expected value of a single roll of a die is $(1 + \cdots + 6)/6 = 3.5$. Therefore, the expected value for $k$ rolls is $3.5k$ and the overall expected value therefore is $\sum_{k=1}^{6} (k + 3.5k)/6 = \sum_{k=1}^{6} 3k/4 = 63/4$ because a $k$ in the first roll occurs with probability $1/6$.

**2.4.2** $m_X'(t) = e^{-r\delta} e^{e^t r\delta} = e^{-r\delta} e^{e^t r\delta} e^t r\delta$ and $m_X''(t) = e^{-r\delta} e^{e^t r\delta} e^t r\delta + e^{-r\delta} e^{e^t r\delta} (e^t r\delta)^2$. Hence $E(X) = m_X'(0) = r\delta$ and $Var(X) = m_X''(0) - (r\delta)^2 = r\delta + (r\delta)^2 - (r\delta)^2 = r\delta$.

**2.4.3** $P(\text{accident}) = 1 - P(\text{no accident}) = 1 - e^{-\delta x}(\delta x)^0/0! = 1 - e^{-1.11x/100,000,000}$

**2.4.4** Since $f(k) = e^{-\delta r}(\delta r)^k/k!$, it follows that

$$P(9 \cdot 10^6 - 3000 \leq k \leq 9 \cdot 10^6 + 3000) = \sum_{k=9 \cdot 10^6 - 3000}^{9 \cdot 10^6 + 3000} \frac{e^{-9 \cdot 10^6}(9 \cdot 10^6)^k}{k!} \approx 0.68.$$

Furthermore, the expectation and variance are both equal to $\delta r = 9,000,000$.

**2.5.9** For $x \in R(X) = [0, X(2)] = [0, 4]$ the inequality $X(t) \leq x$ is satisfied if and only if $0 \leq s^2 - 4s + x$ or, equivalently, if and only if $x \in [0, 2 - \sqrt{4 - x}] \cup [2 + \sqrt{4 - x}, 4]$. Hence

$$F(x) = P(X \leq x) = \begin{cases} 0 & \text{if } x < 0, \\ 2(2 - \sqrt{4 - x})/4 = 1 - (1/2)\sqrt{4 - x} & \text{if } x \in [0, 4], \\ 1 & \text{if } x > 4 \end{cases}$$

and

$$f(x) = F'(x) = \begin{cases} 1/(4\sqrt{4 - x}) & \text{if } x \in [0, 4], \\ 0 & \text{otherwise.} \end{cases}$$

This yields

$$E(X) = \int_{-\infty}^{\infty} x f(x)\, dx = \int_0^4 \frac{x}{4\sqrt{4 - x}}\, dx = \frac{8}{3}$$

and

$$E(X^2) = \int_{-\infty}^{\infty} x^2 f(x)\, dx = \int_0^4 \frac{x^2}{4\sqrt{4 - x}}\, dx = \frac{128}{15},$$

and, by implication, $\text{Var}(X) = 128/15 - (8/3)^2 = 64/45$.

**2.5.10** $E(X) = \int_S X(s) p(s)\, ds = \int_0^4 s(4 - s)\, ds/4 = 8/3$ and $E(X^2) = \int_S X(s)^2 p(s)\, ds = \int_0^4 s^2(4 - s)^2\, dt/4 = 128/15$, as desired.

**2.5.12 a)** $1 = \int_1^3 C x^2\, dx = 26C/3$. Hence $C = 3/26$.

**b)**

$$F(x) = \begin{cases} 0 & \text{if } x < 1, \\ (3/26) \int_1^x f(t)\, dt = (x^3 - 1)/26 & \text{if } x \in [1, 3], \\ 1 & \text{if } x > 3. \end{cases}$$

**c)** $P(1 \leq X \leq 2) = F(2) - F(1) = 7/26$

**2.5.14** $1 = \int_{-\infty}^{\infty} f(x)\, dx = \int_0^{\infty} 2e^{-\alpha x}\, dx = 2/\alpha$. Hence $\alpha = 2$ and

$$F(x) = \begin{cases} 0 & \text{if } x \leq 0, \\ \int_0^x 2e^{-2t}\, dt = 1 - e^{-2x} & \text{if } x > 0. \end{cases}$$

**2.5.15**

$$F(x) = \begin{cases} 0 & \text{if } x < 0, \\ \left(2\sqrt{x/2} + \int_{\sqrt{x/2}}^2 x/s^2\, ds\right)/4 = \sqrt{x/2} - x/8 & \text{if } x \in [0, 8], \\ 1 & \text{if } x > 8. \end{cases}$$

Hence

$$f(x) = F'(x) = \begin{cases} 1/(2\sqrt{2x}) - 1/8 & \text{if } x \in [0, 8], \\ 0 & \text{otherwise.} \end{cases}$$

**2.5.16 a)** $E(X) = \int_0^2 \int_0^2 s^2 t \, ds \, dt / 4 = 4/3$

**b)** $E(X) = \int_0^8 x(1/(2\sqrt{2x}) - 1/8) \, dx = 16/3 - 4 = 4/3$

**2.5.18 a)** $1 = C \int_1^3 1/x \, dx = C \ln(3)$, and therefore, $C = 1/\ln(3)$.

**b)** $P(X > 2) = \left( \int_2^3 1/x \, dx \right) / \ln(3) = (\ln(3) - \ln(2))/\ln(3) = 1 - \ln(2)/\ln(3)$

**c)** $E(X) = \left( \int_1^3 dx \right) / \ln(3) = 2/\ln(3)$ and $E(X^2) = \left( \int_1^3 x \, dx \right) / \ln(3) = 4/\ln(3)$. Hence $\text{Var}(X) = 4(\ln(3) - 1)/\ln^2(3)$.

**2.5.20** $1 = \lambda \int_{-\infty}^{\infty} e^{-|s|} \, ds = 2\lambda$, and therefore, $\lambda = 1/2$. Hence $E(X) = \int_{-\infty}^{\infty} |s| e^{-|s|} \, ds/2 = \int_0^{\infty} se^{-s} \, ds = 1$, $E(X^2) = \int_{-\infty}^{\infty} |s|^2 e^{-|s|} \, ds/2 = \int_0^{\infty} s^2 e^{-s} \, ds = 2$ and $\text{Var}(X) = 2 - 1^2 = 1$.

**2.5.21** Since $|s| \leq x$ if and only if $-x \leq s \leq x$, it follows that

$$F(x) = \begin{cases} \int_{-x}^x e^{-|s|}/2 \, ds = \int_0^x e^{-s} \, ds = 1 - e^{-x} & \text{if } x \geq 0, \\ 0 & \text{otherwise.} \end{cases}$$

**2.5.22** $1 = \lambda \int_{-\infty}^{\infty} \int_{-\infty}^{\infty} e^{-\sqrt{s^2+t^2}} \, ds dt = \lambda \int_0^{2\pi} \int_0^{\infty} e^{-r} r \, dr d\theta = 2\pi\lambda$, and therefore, $\lambda = 1/(2\pi)$. Hence

$$E(X) = \frac{1}{2\pi} \int_{-\infty}^{\infty} \int_{-\infty}^{\infty} \sqrt{s^2 + t^2} \, e^{-\sqrt{s^2+t^2}} \, ds dt = \frac{1}{2\pi} \int_0^{2\pi} \int_0^{\infty} r^2 e^{-r} \, dr d\theta = 2$$

and, similarly,

$$E(X^2) = \frac{1}{2\pi} \int_0^{2\pi} \int_0^{\infty} r^3 e^{-r} \, dr d\theta = 6.$$

Thus $\text{Var}(X) = 6 - 2^2 = 2$.

**2.5.23**

$$F(x) = \begin{cases} 0 & \text{if } x < 0, \\ \int_0^{2\pi} \int_0^x e^{-r} r \, dr d\theta / 2\pi = 1 - (1 + x)e^{-x} & \text{if } x \geq 0. \end{cases}$$

**2.6.12** You need to evaluate $\int_0^{\sqrt{2}} \int_0^{\pi/2} 9e^{-9r^2/8} r \, d\theta \, dr / (8\pi)$.

**2.6.13** Since $\sqrt{2\pi} f'(x) = -(x - \mu)e^{-(x-\mu)^2/2\sigma^2}/\sigma^2$ and $\sqrt{2\pi} f''(x) = (x-\mu)^2 e^{-(x-\mu)^2/2\sigma^2}/\sigma^4 - e^{-(x-\mu)^2/2\sigma^2}/\sigma^2$, it follows that $f'(x) = 0$ at $x = \mu$ and $f''(x) = 0$ at $x = \mu \pm \sigma$.

**2.6.14** Since the $z$-scores of 67.5 and 75 are $-1$ and 2, respectively, it follows that the percentage in question is approximately $68 + (95 - 68)/2 = 81.5\%$.

**2.6.16** The $z$-scores are $(12 - 105)/15 = 1$ and $(115 - 95)/10 = 2$, respectively. Thus the approximate corresponding percentiles are $50 + 34 = 84\%$ and $50 + 47.5 = 97.5\%$.

**2.6.17 a)** 0.95, **c)** 0.815, **e)** 0.84

**2.6.19** Use a computer to evaluate the following integrals: **a)** $\int_{-1/2}^{1/2} z(x)\, dx$, **b)** $1 - \int_{-4/5}^{4/5} z(x)\, dx$, **c)** $\int_{3/5}^{3/2} z(x)\, dx$, **d)** $\int_{-\infty}^{-13/10} z(x)\, dx$

**2.7.4 a)** $\mu = 10000 \cdot 0.2 = 2000$ and $\sigma^2 = 10000 \cdot 0.2 \cdot 0.8 = 1600$. Hence

$$P(1940 \le X \le 1980) \approx \int_{1939.5}^{1980.5} \frac{e^{-(x-2000)^2/(2\cdot 40)}}{40\sqrt{2\pi}}\, dx$$

$$= \int_{-60.5/40}^{-19.5/40} \frac{e^{-x^2/2}}{\sqrt{2\pi}}\, dx \approx 0.2477.$$

**b)** $\sum_{k=1940}^{1980} \binom{10000}{k} 0.2^k 0.8^{10000-k} \approx 0.2488$

**2.8.19**

$$F(x,y) = \begin{cases} 0 & \text{if } x < 0 \text{ or } y < 0, \\ \sqrt{x} & \text{if } (x \in [0,1] \text{ and } y > 1) \text{ or } (x,y \in [0,1] \text{ and } \sqrt{x} < y), \\ y & \text{if } (y \in [0,1] \text{ and } x > 1) \text{ or } (x,y \in [0,1] \text{ and } \sqrt{x} \ge y), \\ 1 & \text{if } x > 1 \text{ and } y > 1. \end{cases}$$

There is no joint density because $\partial^2 F/\partial x \partial y = 0$.

**2.8.21**

$$F(x,y) = \begin{cases} 0 & \text{if } x < 0 \text{ or } y < 0, \\ \int_{-\sqrt{x}}^{\sqrt{x}} \int_{-\sqrt{y}}^{\sqrt{y}} p(s,t)\, dt ds & \text{if } x,y \ge 0, \end{cases}$$

and

$$f(x,y) = \frac{\partial^2}{\partial x \partial y} F(x,y)$$

$$= \begin{cases} 0 & \text{if } x < 0 \text{ or } y < 0, \\ p(\sqrt{x}, \sqrt{y})/\sqrt{xy} = e^{-(x+y)/2}/(2\pi\sqrt{xy}) & \text{if } x,y \ge 0. \end{cases}$$

**2.8.22** $E(\sqrt{XY}) = \int_{-\infty}^{\infty} \int_{-\infty}^{\infty} e^{-(x+y)/2}/2\pi\, dxdy = 2/\pi$

**2.8.23** $E(\sqrt{XY}) = \int_{-\infty}^{\infty} \int_{-\infty}^{\infty} |st| e^{-(s^2+t^2)/2}/(2\pi)\, dsdt = 4\int_0^{\infty} \int_0^{\infty} ste^{-(s^2+t^2)/2}/(2\pi)\, dsdt = 2/\pi$

**2.8.24** $f(1,3) = 1/3$, $f(1,1) = 1/6$, $f(1,4) = 0$, $f(2,3) = 0$, $f(2,1) = 1/4$, $f(2,4) = 1/4$, and $f(x,y) = 0$ otherwise.

**2.8.25** Since

$$f(i,j) := \begin{cases} 2/(n(n+1)) & \text{if } i,j \in \{1,\dots,n\} \text{ and } j \le i, \\ 0 & \text{otherwise,} \end{cases}$$

and therefore,

$$f_X(i) = \frac{2i}{n(n+1)}$$

and

$$f_Y(j) = \frac{2(n-j+1)}{n(n+1)}$$

for all $i, j \in \{1, \ldots, n\}$. Furthermore,

$$E(X) = \sum_{i=1}^{n} i f_X(i) = \sum_{i=1}^{n} \frac{2i^2}{n(n+1)} = \frac{2n+1}{3}$$

and

$$E(Y) = \sum_{j=1}^{n} j f_Y(j) = \sum_{j=1}^{n} \frac{2j(n-j+1)}{n(n+1)} = n+1 - \frac{2n+1}{3} = \frac{n+2}{3}.$$

**2.8.26** Since

$$1 = \int_{-\infty}^{\infty} \int_{-\infty}^{\infty} f(x,y) \, dx \, dy = \int_{0}^{\infty} \int_{0}^{\infty} C e^{-x-2y} \, dx \, dy = \frac{C}{2},$$

we may infer that $C = 2$. Hence

$$f_X(x) = \int_{-\infty}^{\infty} f(x,y) \, dy = \begin{cases} e^{-x} \int_0^\infty 2e^{-2y} \, dy = e^{-x} & \text{if } x \geq 0, \\ 0 & \text{otherwise} \end{cases}$$

and

$$f_Y(y) = \int_{-\infty}^{\infty} f(x,y) \, dx = \begin{cases} 2e^{-2y} \int_0^\infty e^{-x} \, dy = 2e^{-2y} & \text{if } y \geq 0, \\ 0 & \text{otherwise.} \end{cases}$$

**2.9.13** They are not independent because $f(i,j) \neq f_X(i) f_Y(j)$.

**2.9.14** $f(i,j) = 1/n^2$ for all $(i,j) \in \{1, \ldots, n\} \times \{1, \ldots, n\}$, and therefore, $f_X(i) = \sum_{i=1}^{n} 1/n^2 = 1/n = f_Y(j)$ for all $i, j \in \{1, \ldots, n\}$. Thus $X$ and $Y$ are independent because $f(i,j) = f_X(i) f_Y(j)$.

**2.9.15** Since all points in $S$ are equally likely to be chosen, it follows that

$$f(x,y) = \begin{cases} 1/2 & \text{if } x, y \in [-1,0] \text{ or } x, y \in [0,1], \\ 0 & \text{otherwise.} \end{cases}$$

Hence

$$f_X(x) = \begin{cases} \int_{-1}^{1} f(x,y) \, dy = 1/2 & \text{if } x \in [-1,1], \\ 0 & \text{otherwise} \end{cases}$$

and

$$f_Y(y) = \begin{cases} \int_{-1}^{1} f(x, y)\, dx = 1/2 & \text{if } y \in [-1, 1], \\ 0 & \text{otherwise.} \end{cases}$$

Consequently, $f(x, y) \neq f_X(x) f_Y(y)$, and therefore, $X$ and $Y$ are not independent.

**2.9.16 a)**

$$f(x, y, z) = f_X(x) f_Y(y) f_Z(z) = \begin{cases} 1 & \text{if } x, y, z \in [0, 1], \\ 0 & \text{otherwise.} \end{cases}$$

**b)**

$$F(u) = P(XYZ \leq u) = \begin{cases} 0 & \text{if } u < 0, \\ u + \int_u^1 u/x\, dx + \int_u^1 \int_{u/x}^1 u/xy\, dy dx & \text{if } u \in [0, 1], \\ 1 & \text{if } u > 1, \end{cases}$$

$$= \begin{cases} 0 & \text{if } u < 0, \\ u(1 - \ln(u) + \ln^2(u)/2) & \text{if } u \in [0, 1], \\ 1 & \text{if } u > 1, \end{cases}$$

and

$$f(u) = F'(u) = \begin{cases} \ln^2(u)/2 & \text{if } u \in [0, 1], \\ 0 & \text{otherwise.} \end{cases}$$

**c)**

$$E(XYZ) = \int_0^1 \frac{u \ln^2(u)}{2}\, du = \frac{1}{8}$$

and

$$E(XYZ) = \int_0^1 \int_0^1 \int_0^1 xyz\, dx dy dz = \frac{1}{8},$$

**2.9.17** The cumulative density of $X^2$ and $Y^2$ is

$$H(x, y) = \begin{cases} 0 & \text{if } x \leq 0 \text{ or } y \leq 0, \\ \int_{-\sqrt{y}}^{\sqrt{y}} \int_{-\sqrt{x}}^{\sqrt{x}} f(u) g(v)\, du dv & \text{if } x > 0 \text{ and } y > 0, \end{cases}$$

and the regular density is

$$h(x, y) = \frac{\partial^2 H}{\partial x \partial y}$$

$$= \begin{cases} 0 & \text{if } x \leq 0 \text{ or } y \leq 0, \\ \dfrac{f(\sqrt{x}) + f(-\sqrt{x})}{2\sqrt{x}} \cdot \dfrac{g(\sqrt{y}) + g(-\sqrt{y})}{2\sqrt{y}} & \text{if } x > 0 \text{ and } y > 0. \end{cases}$$

Consequently, for $x > 0$ we find that

$$
\begin{aligned}
h_{X^2}(x) &= \frac{f(\sqrt{x}) + f(-\sqrt{x})}{2\sqrt{x}} \int_0^\infty \frac{g(\sqrt{y}) + g(-\sqrt{y})}{2\sqrt{y}}\, dy \\
&= \frac{f(\sqrt{x}) + f(-\sqrt{x})}{2\sqrt{x}} \int_0^\infty (g(u) + g(-u))\, du \\
&= \frac{f(\sqrt{x}) + f(-\sqrt{x})}{2\sqrt{x}} \int_{-\infty}^\infty g(u)\, du = \frac{f(\sqrt{x}) + f(-\sqrt{x})}{2\sqrt{x}}.
\end{aligned}
$$

Hence

$$
h_{X^2}(x) = \begin{cases} 0 & \text{if } x \le 0, \\ \dfrac{f(\sqrt{x}) + f(-\sqrt{x})}{2\sqrt{x}} & \text{if } x > 0, \end{cases}
$$

and similarly,

$$
h_{Y^2}(y) = \begin{cases} 0 & \text{if } y \le 0, \\ \dfrac{g(\sqrt{y}) + g(-\sqrt{y})}{2\sqrt{y}} & \text{if } y > 0. \end{cases}
$$

This shows that $X^2$ and $Y^2$ are independent because $h(x,y) = h_{X^2}(x)h_{Y^2}(y)$.

**2.9.18** Since $X^2$ and $Y^2$ are independent (by Exercise 2.9.17), we find that

$$
\begin{aligned}
\mathrm{Var}(XY) &= E(X^2 Y^2) - E(XY)^2 = E(X^2)E(Y^2) - E(X)^2 E(Y)^2 \\
&= (\sigma_X^2 + \mu_X^2)(\sigma_Y^2 + \mu_Y^2) - \mu_X^2 \mu_Y^2 \\
&= \sigma_X^2 \sigma_Y^2 + \sigma_X^2 \mu_Y^2 + \sigma_Y^2 \mu_X^2.
\end{aligned}
$$

**2.9.19** See the proof of Lemma 3.2.5.

**2.9.20** Since $2X - 3Y$ is normal with mean $\mu = 2\mu_X - 3\mu_Y = 1$ and variance $\sigma^2 = 4\sigma_X^2 + 9\sigma_Y^2 = 100$, it follows that

$$
P(|2X - 3Y| < 10) = \int_{-10}^{10} \frac{e^{-(x-1)^2/200}}{10\sqrt{2\pi}}\, dx = \int_{-11/10}^{9/10} \frac{e^{-x^2/2}}{\sqrt{2\pi}}\, dx \approx 0.6803.
$$

**2.9.22** $E(\sum_{k=1}^n \lambda_k X_k) = \sum_{k=1}^n \lambda_k E(X_k) = \mu \sum_{k=1}^n \lambda_k$ and
$\mathrm{Var}(\sum_{k=1}^n \lambda_k X_k) = \sum_{k=1}^n \lambda_k^2 \mathrm{Var}(X_k)$, and therefore, the standard deviation is $\sigma \sqrt{\sum_{k=1}^n \lambda_k^2}$.

**2.9.23** Given the assumption of independence, the joint density is

$$
f(x,y) = \frac{e^{-x^2/32}}{4\sqrt{2\pi}} \cdot \frac{e^{-y^2/32}}{4\sqrt{2\pi}} = \frac{e^{-(x^2+y^2)/32}}{32\pi},
$$

and therefore,

$$P(X^2 + Y^2 \leq 1) = \int_0^{2\pi} \int_0^1 \frac{e^{-r^2/32}r}{32\pi} \, dr d\theta$$

$$= \frac{1}{16} \int_0^1 e^{-r^2/32}r \, dr = -e^{-r^2/32} \Big|_0^1 = 1 - e^{-1/32}.$$

**2.9.24** Given the assumption of independence, it follows that $X - Y$ is normal with $E(X - Y) = E(X) - E(Y) = 0$ and $\sigma^2 = \text{Var}(X - Y) = \text{Var}(X) + (-1)^2 \text{Var}(Y) = 9$. Hence

$$P(X - Y \leq 6) = P(X - Y \leq 2\sigma) \approx 0.5 + \frac{0.95}{2} = 0.975.$$

**2.9.27 a)** $E(X) = \int_0^1 s \, ds = 1/2$, $E(Y) = \int_0^1 (s - s^2) \, ds = 1/6$, and $E(XY) = \int_0^1 s(s - s^2) \, ds = 1/12$.
**b)** According to a) we have $E(XY) = E(X)E(Y)$, but $X$ and $Y$ are nonetheless not independent because

$$P(\{X \in [0, 2/9]\}) P(\{Y \in [0, 2/9]\}) = P([0, 2/9]) P([0, 1/3] \cup [2/3, 1]) = \frac{4}{27}$$

$$\neq \frac{2}{9} = P([0, 2/9] \cap ([0, 1/3] \cup [2/3, 1])) = P(\{X \in [0, 2/9]\} \cap \{Y \in [0, 2/9]\})$$

**2.10.1 a)** $E(w) = 0$. This is intuitively obvious, and it also follows from the fact that

$$f(w) = \frac{\sqrt{m}e^{-mw^2/2kT}}{\sqrt{2\pi kT}}$$

is a normal density with mean $\mu = 0$. Furthermore,

$$E(|w|) = \int_{-\infty}^{\infty} |w| f(w) \, dw = 2 \int_0^{\infty} \frac{w\sqrt{m}e^{-mw^2/2kT}}{\sqrt{2\pi kT}} \, dw = \frac{\sqrt{2kT}}{\sqrt{\pi m}}.$$

**b)** $E(u^2 + v^2 + w^2) = 3E(w^2) = 3\sigma^2 = 3kT/m$
**c)**

$$E\left(\sqrt{u^2 + v^2 + w^2}\right) = \int_0^{2\pi} \int_0^{\pi} \int_0^{\infty} \frac{\sqrt{m}^3 \rho e^{-m\rho^2/2kT}}{\sqrt{2\pi kT}^3} \rho^2 \sin(\phi) d\rho d\phi d\theta$$

$$= \frac{4\pi\sqrt{m}^3}{\sqrt{2\pi kT}^3} \int_0^{\infty} \rho^3 e^{-m\rho^2/2kT} d\rho$$

$$= \frac{\sqrt{m}^3}{\sqrt{2\pi}\sqrt{kT}^3} \int_0^{\infty} ue^{-mu/2kT} du$$

$$= \frac{\sqrt{m}^3}{\sqrt{2\pi}\sqrt{kT}^3} \cdot \frac{4k^2T^2}{m^2} = \frac{\sqrt{8kT}}{\sqrt{\pi m}}$$

**2.11.16 a)** $E(X) = 1/2$, $E(Y) = 1/3$, $E(XY) = 1/4$, $E(X^2) = 1/3$, $E(Y^2) = 1/5$, and therefore, $\rho(X, Y) = \sqrt{15}/4 \approx 0.968$.
**b)** $E(X) = 0$, $E(Y) = 1/3$, $E(XY) = 0$, and therefore, $\rho(X, Y) = 0$.
**c)** $E(X) = 3 \int_0^1 s \cdot s^2 \, ds = 3/4$, $E(Y) = 3/5$, $E(XY) = 1/2$, $E(X^2) = 3/5$, $E(Y^2) = 3/7$, and therefore, $\rho(X, Y) = \sqrt{35}/6 \approx 0.986$.

**2.11.20** $Cov(X, Y) = 0$ for a) and c) because $X$ and $Y$ are independent (and this is so because in either case we have $f(x, y) = f_X(x) f_Y(y)$).
**b)** $E(X) = 2 \int_0^1 \int_0^x x \, dy \, dx = 2/3$, $E(Y) = 1/3$, $E(XY) = 1/4$, and therefore, $Cov(X, Y) = 1/36$.

**2.11.21** The correlation is largest for the curve on the left and smallest for the curve in the middle.

**2.11.22** If $X(1) := X(2) := 1$, $X(3) := X(4) := -1$, $Y(1) := Y(3) := 1$, and $Y(2) := Y(4) := -1$, then $|X| = |Y| = 1$, $\sigma(X) = \sigma(Y) = 1$, and $\rho(X, Y) = 0$ because $X$ and $Y$ are independent.

**2.11.23** $E(X) = E(Y_n) = 0$ and if $n > 1$, then

$$E(XY_n) = \frac{1}{2\pi} \int_0^{2\pi} \sin(s) \sin(ns) \, ds$$
$$= \frac{1}{2\pi} \int_0^{2\pi} \frac{\cos((n-1)s) - \cos((n+1)s)}{2} \, ds = 0,$$

and therefore, $Cov(X, Y_n) = 0$. However, $X$ and $Y_n$ are not independent, because if $n$ is odd, then

$$P(\{X \in [0, \sin(\pi/(3n))]\}) P(\{Y_n \in [0, \sin(\pi/(3n))]\}) = \frac{1}{9n^2}$$
$$\neq \frac{1}{3n^2} = P(\{X \in [0, \sin(\pi/(3n))]\} \cap \{Y_n \in [0, \sin(\pi/(3n))]\}),$$

and if $n$ is even, then

$$P(\{X \in [0, \sin(\pi/(3n))]\}) P(\{Y_n \in [0, \sin(\pi/(3n))]\}) = \frac{1}{9n^2}$$
$$\neq \frac{1}{6n^2} = P(\{X \in [0, \sin(\pi/(3n))]\} \cap \{Y_n \in [0, \sin(\pi/(3n))]\}).$$

**2.11.24 a)** The curve described by $c_\alpha$ is an ellipse with major radius $a = 3$ and minor radius $b = 1$ that is rotated counterclockwise by $\alpha$. That is to say, $\alpha$ is the angle between the positive $x$-axis and the major radius line of the ellipse.

**b)** $E(X_\alpha) = E(Y_\alpha) = 0$,

$$E(X_\alpha^2) = \frac{8\cos^2(\alpha) + 1}{2},$$

$$E(Y_\alpha^2) = \frac{8\sin^2(\alpha) + 1}{2},$$

$$E(X_\alpha Y_\alpha) = 4\sin(\alpha)\cos(\alpha) = 2\sin(2\alpha),$$

and therefore

$$R(\alpha) = \frac{4\sin(2\alpha)}{\sqrt{(8\cos^2(\alpha) + 1)(8\sin^2(\alpha) + 1)}} = \frac{4\sin(2\alpha)}{\sqrt{16\sin^2(2\alpha) + 9}}.$$

**c)** Use a graphing calculator or a computer.

**d)** Setting $x = \sin(2\alpha)$, we need to find the maxima and minima of the function

$$f(x) := \frac{4x}{\sqrt{16x^2 + 9}}$$

on the interval $[-1, 1]$. Taking the derivative of $f$, we readily find that $f'(x) > 0$ for all $x \in \mathbb{R}$. Hence $f$ is strictly increasing, and therefore, its maximum and minimum on $[-1, 1]$ are $f(1) = 4/5 = 0.8$ and $f(-1) = -0.8$, respectively.

# Chapter 3

**3.1.3** $P(|X_{40}/40 - 1/4| \leq 0.1) = \sum_{k=6}^{14} \binom{40}{k}(1/4)^k(3/4)^{40-k} \approx 0.902289$,
$P(|X_{400}/400 - 1/4| \leq 0.1) = \sum_{k=60}^{140} \binom{400}{k}(1/4)^k(3/4)^{400-k} \approx 0.999996$, and
$P(|X_{4000}/4000 - 1/4| \leq 0.1) = \sum_{k=600}^{1400} \binom{4000}{k}(1/4)^k(3/4)^{4000-k}$
$\approx \left(\sum_{k=1}^{44} 9/10^k\right) + 8/10^{45} \approx 1$. The values are rapidly increasing to 1, and the third value is essentially indistinguishable from 1.

**3.2.8** If $0 < x \leq 1$, then $1 - 1/x \leq 0$, and therefore, $\sqrt{1 - e^{-x/2}} > 0 \geq 1 - 1/x$. So we only need to consider the case $x > 1$, or equivalently, $1 - 1/x > 0$. Given this inequality, we find that

$$\sqrt{1 - e^{-x/2}} > 1 - \frac{1}{x} \Leftrightarrow 1 - e^{-x/2} > 1 - \frac{2}{x} + \frac{1}{x^2} \Leftrightarrow \frac{1}{e^{x/2}} < \frac{2}{x} - \frac{1}{x^2}.$$

Since $e^x$ is concave up and since the tangent line at $x = 0$ is $y = x+1$, it follows that $e^x \geq 1 + x$ for all $x \in \mathbb{R}$. Thus $1/e^{x/2} \leq 1/(1 + x/2)$, and consequently, it only remains to be shown that $1/(1 + x/2) < 2/x - 1/x^2$ for all $x > 1$. Using elementary algebra, we may infer that

$$\frac{1}{1 + x/2} < \frac{2}{x} - \frac{1}{x^2} \Leftrightarrow x^2 < (2x - 1)(1 + x/2) \Leftrightarrow 1 < 3x/2,$$

and the last of these inequalities is trivially satisfied because $x > 1$.

**3.2.9** Setting $X := \sum_{k=1}^{n} X_k/n$, it follows that $E(X) = \mu$ and $\sigma^2 = \text{Var}(X) = \sum_{k=1}^{n} \sigma_k^2/n^2$. Consequently, Chebyshev's inequality implies that

$$P\left(\left|\frac{1}{n}\sum_{k=1}^{n} X_k - \mu\right| \leq \varepsilon\right) = P(|X - \mu| \leq \varepsilon) \geq 1 - \frac{\sigma^2}{\varepsilon^2} = 1 - \frac{1}{\varepsilon^2 n^2}\sum_{k=1}^{n}\sigma_k^2,$$

as desired.

**3.2.10** Let $\varepsilon$ be a given positive number and assume that $(X_k)_{k=1}^{\infty}$ is a sequence of independent random variables with common mean $\mu$ such that $\sigma_k \leq \sqrt{k}^{\alpha}$ for some $\alpha \in (0, 1)$ and all positive integers $k$. Using the result of Exercise 3.2.9, it follows that

$$1 \geq \lim_{n\to\infty} P\left(\left|\frac{1}{n}\sum_{k=1}^{n} X_k - \mu\right| \leq \varepsilon\right) \geq \lim_{n\to\infty}\left(1 - \frac{1}{\varepsilon^2 n^2}\sum_{k=1}^{n}\sigma_k^2\right)$$

$$\geq \lim_{n\to\infty}\left(1 - \frac{1}{\varepsilon^2 n^2}\sum_{k=1}^{n} k^{\alpha}\right) \geq \lim_{n\to\infty}\left(1 - \frac{1}{\varepsilon^2 n^2}\int_{1}^{n+1} x^{\alpha}\,dx\right)$$

$$\geq \lim_{n\to\infty}\left(1 - \frac{(n+1)^{\alpha+1} - 1}{(\alpha+1)\varepsilon^2 n^2}\right) = 1,$$

and therefore,

$$\lim_{n\to\infty} P\left(\left|\frac{1}{n}\sum_{k=1}^{n} X_k - \mu\right| \leq \varepsilon\right) = 1,$$

as desired.

**3.2.11** Since the function $x/\ln(x)$ is increasing on the interval $[e, \infty)$, it follows that for any fixed value $m \geq 3$ we have

$$1 \geq \lim_{n\to\infty} P\left(\left|\frac{1}{n}\sum_{k=1}^{n} X_k - \mu\right| \leq \varepsilon\right) \geq \lim_{n\to\infty}\left(1 - \frac{1}{\varepsilon^2 n^2}\sum_{k=1}^{n}\sigma_k^2\right)$$

$$= \lim_{n\to\infty}\left(1 - \frac{1}{\varepsilon^2 n^2}\sum_{k=m}^{n}\sigma_k^2\right) \geq \lim_{n\to\infty}\left(1 - \frac{1}{\varepsilon^2 n^2}\sum_{k=m}^{n}\frac{k}{\ln(k)}\right)$$

$$\geq \lim_{n\to\infty}\left(1 - \frac{1}{\varepsilon^2 n^2}\int_{m}^{n+1}\frac{x}{\ln(x)}\,dx\right) \geq \lim_{n\to\infty}\left(1 - \frac{1}{\varepsilon^2 n^2 \ln(m)}\int_{m}^{n+1} x\,dx\right)$$

$$\geq \lim_{n\to\infty}\left(1 - \frac{(n+1)^2}{2\varepsilon^2 n^2 \ln(m)}\right) = 1 - \frac{1}{2\varepsilon^2 \ln(m)}.$$

Since this inequality is valid for all $m \geq 3$, we may infer that

$$1 \geq \lim_{n\to\infty} P\left(\left|\frac{1}{n}\sum_{k=1}^{n} X_k - \mu\right| \leq \varepsilon\right) \geq 1,$$

and therefore,

$$\lim_{n\to\infty} P\left(\left|\frac{1}{n}\sum_{k=1}^{n} X_k - \mu\right| \leq \varepsilon\right) = 1,$$

as desired.

**3.2.14** For $n = 1,000,000 = 10^6$ we find that $P(|\frac{1}{n}\sum_{k=1}^{n} X_k - 2| > 0.003) \leq \sigma^2/(10^6 \cdot 0.003^2) \leq 1/9 \approx 0.11$, and therefore, the outcome is not very surprising. However for $n = 1,000,000,000 = 10^9$, the same type of estimate yields $P(|\frac{1}{n}\sum_{k=1}^{n} X_k - 2| > 0.003) \leq \sigma^2/(10^9 \cdot 0.003^2) \leq 1/9000 \approx 0.00011$, and since a chance of 1 in 9000 may rightly be said to be very small, the outcome would be very surprising.

**3.2.15** Using Proposition 3.2.6, we find that

$$P\left(\left|\frac{1}{n}\sum_{k=1}^{n} X_k - 2\right| > 0.003\right) \leq 1 - \sqrt{1 - e^{-10^6 \cdot 0.003^2/2\sigma^2}}$$

$$\leq 1 - \sqrt{1 - e^{-9/2}} \approx 0.0056,$$

and therefore, the outcome would be quite surprising.

**3.3.10** We need to show that

$$P(\{X_1 \in B_1\} \cap \cdots \cap \{X_n \in B_n\}) = P(\{X_1 \in B_1\})\ldots P(\{X_n \in B_n\})$$

for all suitable sets $B_1, \ldots, B_n \subset S$. Since $R(X_k) = \{0,1\}$ for all $k \in \{1, \ldots, n\}$, it follows that $\{X_k \in B_k\} = \emptyset$ whenever $B_k \cap \{0,1\} = \emptyset$. So if there is a $k \in \{1, \ldots, n\}$ such that $B_k \cap \{0,1\} = \emptyset$, then both sides of the equation above are zero and therefore equal to each other. Furthermore, if $\{0,1\} \subset B_k$, then $\{X_k \in B_k\} = S$, and by implication, the set $\{X_k \in B_k\}$ can be eliminated from both sides of the equation above. Consequently, what we really need to show is this: if $I \subset \{1, \ldots, n\}$ is a set with at least two elements such that $\{X_k \in B_k\} \cap \{0,1\}$ is equal to either $\{0\}$ or $\{1\}$, then

$$P\left(\bigcap_{k\in I}\{X_k \in B_k\}\right) = \prod_{k\in I} P(\{X_k \in B_k\}).$$

But given the way the probability measure $P$ on $S$ is defined, this latter statement is equivalent to the following assertion: if $F_k = E_k$ or $F_k = [0,1] \smallsetminus E_k$ for all $k \in \{1, \ldots, n\}$ and if $I \subset \{1, \ldots, n\}$ is a set with at least two elements, then

$$P\left(\bigcap_{k\in I} F_k\right) = \prod_{k\in I} P(F_k) = \frac{1}{2^{\#I}}.$$

The reason why this equation is valid is best understood by looking at an example. Consider for instance the case where $n$ equals 4. Then

$$E_1 = [0, 1/2]$$
$$E_2 = [0, 1/4] \cup [1/2, 3/4]$$
$$E_3 = [0, 1/8] \cup [1/4, 3/8] \cup [1/2, 5/8] \cup [3/4, 7/8]$$
$$E_4 = [0, 1/16] \cup [1/8, 3/16] \cup [1/4, 5/16] \cup [3/8, 7/16]$$
$$\cup [1/2, 9/16] \cup [5/8, 11/16] \cup [3/4, 13/16] \cup [7/8, 15/16],$$

and visually, these sets may be represented as follows:

| | | | | | | | | | | | | | | | | |
|---|---|---|---|---|---|---|---|---|---|---|---|---|---|---|---|---|
| $E_1$: | 1 | 1 | 1 | 1 | 1 | 1 | 1 | 1 | 0 | 0 | 0 | 0 | 0 | 0 | 0 | 0 |
| $E_2$: | 1 | 1 | 1 | 1 | 0 | 0 | 0 | 0 | 1 | 1 | 1 | 1 | 0 | 0 | 0 | 0 |
| $E_3$: | 1 | 1 | 0 | 0 | 1 | 1 | 0 | 0 | 1 | 1 | 0 | 0 | 1 | 1 | 0 | 0 |
| $E_4$: | 1 | 0 | 1 | 0 | 1 | 0 | 1 | 0 | 1 | 0 | 1 | 0 | 1 | 0 | 1 | 0 |

Thus the intersection $E_1 \cap E_2 \cap E_3 \cap E_4$ is represented by the sequence

$$1000000000000000,$$

and its probability therefore is $1/2^4$, as desired. Morever, if we replace, say, $E_2$ and $E_4$ by their complements, so that $F_1 = E_1$, $F_2 = [0, 1] \setminus E_2$, $F_3 = E_3$, and $F_4 = [0, 1] \setminus E_4$, then the corresponding array is

| | | | | | | | | | | | | | | | | |
|---|---|---|---|---|---|---|---|---|---|---|---|---|---|---|---|---|
| $F_1$: | 1 | 1 | 1 | 1 | 1 | 1 | 1 | 1 | 0 | 0 | 0 | 0 | 0 | 0 | 0 | 0 |
| $F_2$: | 0 | 0 | 0 | 0 | 1 | 1 | 1 | 1 | 0 | 0 | 0 | 0 | 1 | 1 | 1 | 1 |
| $F_3$: | 1 | 1 | 0 | 0 | 1 | 1 | 0 | 0 | 1 | 1 | 0 | 0 | 1 | 1 | 0 | 0 |
| $F_4$: | 0 | 1 | 0 | 1 | 0 | 1 | 0 | 1 | 0 | 1 | 0 | 1 | 0 | 1 | 0 | 1 |

and the intersection $F_1 \cap F_2 \cap F_3 \cap F_4$ is represented by the sequence

$$0000010000000000$$

which again has probability $1/2^4$. Finally, if we remove, say, $F_3$ so as to replace $\{1, 2, 3, 4\}$ by $I = \{1, 2, 4\}$, then the corresponding array is

| | | | | | | | | | | | | | | | | |
|---|---|---|---|---|---|---|---|---|---|---|---|---|---|---|---|---|
| $F_1$: | 1 | 1 | 1 | 1 | 1 | 1 | 1 | 1 | 0 | 0 | 0 | 0 | 0 | 0 | 0 | 0 |
| $F_2$: | 0 | 0 | 0 | 0 | 1 | 1 | 1 | 1 | 0 | 0 | 0 | 0 | 1 | 1 | 1 | 1 |
| $F_4$: | 0 | 1 | 0 | 1 | 0 | 1 | 0 | 1 | 0 | 1 | 0 | 1 | 0 | 1 | 0 | 1 |

and the intersection $F_1 \cap F_2 \cap F_4$ is represented by

$$0000000100000001$$

which has probability $1/2^3$ as it should. In generalizing this line of argument, we readily find that the random variables $X_1, \ldots, X_n$ are indeed independent.

**3.3.11** Setting $I := \{x \in \mathbb{R} \mid f(x) \neq 0\}$, it follows that

$$P(X_k \in B) = \sum_{x \in I \cap B} f(x)$$

for all $k \in \{1, \ldots, n\}$, and since the right-hand side of this equation does not depend on $k$, we may infer that

$$P(X_1 \in B) = \cdots = P(X_n \in B),$$

as desired.

**3.3.12** Let

$$X(s) := \begin{cases} 1 & \text{if } s \in [0, 1/2], \\ -1 & \text{if } s \in (1/2, 1] \end{cases}$$

and

$$Y(s) := \begin{cases} \sqrt{2} & \text{if } s \in [0, 1/4], \\ -\sqrt{2} & \text{if } s \in (1/4, 1/2] \\ 0 & \text{if } s \in (1/2, 1]. \end{cases}$$

Then $E(X) = E(Y) = 0$ and $\text{Var}(X) = \text{Var}(Y) = 1$, but the discrete densities of $X$ and $Y$ are obviously not the same because $R(X) \cap R(Y) = \emptyset$.

**3.3.13** Let $F_1 := E_1$ and $F_n := E_n \setminus E_{n-1}$ for all $n \in \mathbb{N} \setminus \{1\}$. Since $E_n \subset E_{n+1}$ for all $n \in \mathbb{N}$, it follows that $F_1 = E_1 \subset E_n$ for all $n \in \mathbb{N}$. Thus $F_1 \cap F_n \subset F_1 \cap E_n \setminus F_1 = \emptyset$ for all $n > 1$. Furthermore, if $m > n > 1$, then $E_n \setminus E_{n-1} \subset E_n \subset E_{m-1}$ (because $E_k \subset E_{k+1}$ for all $k \in \mathbb{N}$), and therefore,

$$F_m \cap F_n = (E_m \setminus E_{m-1}) \cap (E_n \setminus E_{n-1}) \subset (E_m \setminus E_{m-1}) \cap E_{m-1} = \emptyset.$$

Consequently, $(F_n)_{n \in \mathbb{N}}$ is a sequence of mutually exclusive events. Moreover, since it is trivially the case that

$$\bigcup_{n=1}^{\infty} F_n = E_1 \cup \bigcup_{n=2}^{\infty} E_n \setminus E_{n-1} \subset \bigcup_{n=1}^{\infty} E_n,$$

it follows that in order to show that

$$\bigcup_{n=1}^{\infty} E_n = \bigcup_{n=1}^{\infty} F_n,$$

we only need to prove that

$$\bigcup_{n=1}^{\infty} E_n \subset \bigcup_{n=1}^{\infty} F_n.$$

So let $s \in \bigcup_{n=1}^{\infty} E_n$ and let $n$ be the smallest positive integer for which $s \in E_n$. If $n = 1$, then $s \in E_1 = F_1$, and if $n > 1$, then the minimality of $n$ with respect to the condition $s \in E_n$ implies that $s \notin E_{n-1}$, and therefore, $s \in E_n \setminus E_{n-1} = F_n$. So in either case we find that $s \in \bigcup_{n=1}^{\infty} F_n$. Consequently, the third axiom of probability allows us to infer that

$$P\left(\bigcup_{n=1}^{\infty} E_n\right) = P\left(\bigcup_{n=1}^{\infty} F_n\right) = \sum_{n=1}^{\infty} P(F_n) = \lim_{n \to \infty} \sum_{k=1}^{n} P(F_k)$$

$$= P(E_1) + \lim_{n \to \infty} \sum_{k=2}^{n} P(E_k \setminus E_{k-1})$$

$$= P(E_1) + \lim_{n \to \infty} \sum_{k=2}^{n} (P(E_k) - P(E_{k-1})) \quad \text{(because } E_{k-1} \subset E_k\text{)}$$

$$= P(E_1) + \lim_{n \to \infty} (P(E_n) - P(E_1)) = \lim_{n \to \infty} P(E_n),$$

as desired.

**3.3.14** Let $\delta > 0$ and let $s \in S$. We need to show that there is an $N \in \mathbb{N}$ such $|X_n(s)| < \delta$ for all $n \geq N$. Since $(X_n)_{n \in \mathbb{N}}$ converges to zero in probability, it follows that

$$\lim_{n \to \infty} P(|X_n| > \delta/2) = 0.$$

Consequently, there exists an $N \in \mathbb{N}$ such that $P(|X_n| > \delta/2) < P(s)$ for all $n \geq N$, and this in turn implies that $s \notin \{r \in S \mid |X_n(r)| > \delta/2\}$ for all $n \geq N$ (because otherwise $P(|X_n| > \delta/2) \geq P(s)$ for some $n \geq N$). Hence $|X_n(s)| \leq \delta/2 < \delta$ for all $n \geq N$, as desired.

**3.3.15** Since $\lim_{n \to \infty} X_n(s) = \lim_{n \to \infty} s^2/n = 0$ for all $s \in \mathbb{R}$, it follows that $(X_n)_{n \in \mathbb{N}}$ converges to zero pointwise (almost) everywhere on $S = \mathbb{R}$. Consequently, Theorem 3.3.9 implies that $(X_n)_{n \in \mathbb{N}}$ converges to zero in probability.

**3.3.16** Let $M := \bigcup_{n=1}^{\infty} \{1/n, \ldots, n/n\}$. Then $M$ is discrete because $M$ is a discrete union of finite sets. Hence

$$P(M) = \sum_{x \in M} P(\{x\}) = \sum_{x \in M} 0 = 0,$$

and since $X_n(s) = 0$ for all $s \in [0, 1] \setminus M$ and all $n \in \mathbb{N}$, it follows that $(X_n)_{n \in \mathbb{N}}$ converges to zero pointwise almost everywhere (because $\lim_{n \to \infty} X_n(s) = 0$ for all $s \in [0, 1] \setminus M$). Thus $(X_n)_{n \in \mathbb{N}}$ converges to zero in probability by Theorem 3.3.9.

**3.4.3** Since $\mu = 3.5$ and $\sigma^2 = (1^2 + \cdots + 6^2)/6 - (7/2)^2 = 35/12$, it follows that

$$P\left(\left|\frac{1}{400}\sum_{n=1}^{400} X_n - 3.5\right| \le 0.1\right) \approx \int_{3.4}^{3.6} \frac{e^{-(x-3.5)^2/(2(35/12)/400)}}{\sqrt{2\pi}\sqrt{(35/12)/400}}\, dx$$

$$= \int_{-4\sqrt{3}/\sqrt{35}}^{4\sqrt{3}/\sqrt{35}} \frac{e^{-x^2/2}}{\sqrt{2\pi}}\, dx \approx 0.7584.$$

# Chapter 4

**4.1.9** $\bar{x} = (1+2+5+2+6+1+3+2+4+2)/10 = 3$ and

$$s^2 = \frac{1}{9}\left(1^2 + 2^2 + 5^2 + 2^2 + 6^2 + 1^2 + 3^2 + 2^2 + 4^2 + 2^2 - 10 \cdot 3^2\right) = \frac{14}{9}.$$

**4.1.11** The sample correlation coefficient is equal to one, and therefore, the given points $(x_1, y_1), \ldots, (x_5, y_5)$ are located on a straight line with positive slope (by Theorem 4.1.7). The equation of the line happens to be $y = 2x + 1$.

**4.2.4** $y = 127x/91 + 65/91$ (see Figure 4.10).

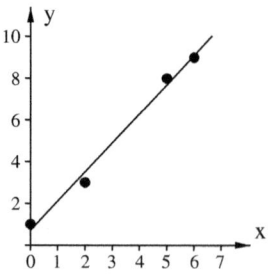

Figure 4.10: graph for Exercise 4.2.4

**4.2.5** Using the result of Exercise 2.11.24, it follows that the equation of the regression line is

$$\frac{y/\sqrt{8\sin^2(\alpha)+1}}{x/\sqrt{8\cos^2(\alpha)+1}} = \frac{4\sin(2\alpha)}{\sqrt{16\sin^2(2\alpha)+9}},$$

or equivalently,

$$y = \frac{4\sin(2\alpha)x}{8\cos^2(\alpha)+1}.$$

**4.2.7** The sample correlation coefficient will decrease to a value closer to $-1$ because, overall, the remaining points will better approximate a line with negative slope.

**4.2.8** The sample correlation coefficient will decrease to a value that is somewhat more distant from 1 because the removal of $A$ causes the remaining points to approximate a line with positive slope somewhat more loosely.

**4.2.10** $E(X) = 1/2$, $E(Y_n) = 1/(n+1)$, $E(X^2) = 1/3$, $E(Y_n^2) = 1/(2n+1)$, $E(XY_n) = 1/(n+2)$, and therefore, $\sigma(X) = 1/\sqrt{12}$,

$$\sigma(Y_n) = \sqrt{\frac{1}{2n+1} - \frac{1}{(n+1)^2}} = \frac{n}{(n+1)\sqrt{2n+1}},$$

$$\mathrm{Cov}(X, Y_n) = \frac{1}{n+2} - \frac{1}{2(n+1)} = \frac{n}{2(n+1)(n+2)},$$

and

$$\rho(X, Y_n) = \frac{\sqrt{3}\sqrt{2n+1}}{n+1}.$$

Consequently, the equation of the regression line is

$$\frac{(n+1)\sqrt{2n+1}(y-1/(n+1))/n}{\sqrt{12}(x-1/2)} = \frac{\sqrt{3}\sqrt{2n+1}}{n+1}$$

or, equivalently,

$$y = \frac{6n(x-1/2)}{(n+1)(n+2)} + \frac{1}{n+1}.$$

Thus the limiting regression line equation, as $n$ tends to infinity, is $y = 0$. In order to see why this is not surprising, we rewrite equation (4.11) in the form

$$\frac{y - E(Y_n)}{x - E(X)} = \frac{\mathrm{Cov}(X, Y_n)}{\sigma(X)^2} = 12(E(XY_n) - E(Y_n)/2)$$

and observe that

$$\lim_{n \to \infty} Y_n(s) = \begin{cases} 0 & \text{if } s \in [0, 1), \\ 1 & \text{if } s = 1. \end{cases}$$

The latter equation shows that $(Y_n)_{n \in \mathbb{N}}$ converges to zero pointwise almost everywhere and therefore the probability for $XY_n$ and $Y_n$ to be ever closer to zero will be ever closer to one as $n$ increases to infinity (see also Theorem 3.3.9). Thus we may expect that $\lim_{n \to \infty} 12(E(XY_n) - E(Y_n)/2) = 0$ and that, by implication, the limiting regression line is $y = 0$.

**4.2.11** Setting

$$\begin{aligned} D(a, b, c) := &(1 - (a - b + c))^2 \\ &+ (-1 - (a + b + c))^2 \\ &+ (0 - (4a + 2b + c))^2 \\ &+ (6 - (25a + 5b + c))^2, \end{aligned}$$

it is easy to see that the equation $\nabla D = \mathbf{0}$ is equivalent to the matrix equation

$$\begin{pmatrix} 643 & 131 & 31 \\ 131 & 31 & 7 \\ 31 & 7 & 4 \end{pmatrix} \begin{pmatrix} a \\ b \\ c \end{pmatrix} = \begin{pmatrix} 150 \\ 28 \\ 6 \end{pmatrix}.$$

Multiplying both sides with the inverse of the matrix on the left yields $a = 5/12$, $b = -49/60$, and $c = -3/10$. Note: the apparent symmetry of the coefficients in the matrix above is not coincidental—think about it.

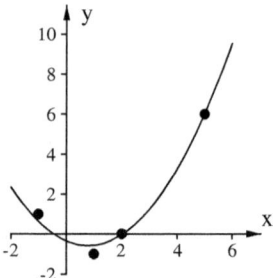

Figure 4.11: graph for Exercise 4.2.11

**4.2.12** The gradient $\nabla D = (\partial D/\partial m, \partial D/\partial b)$ is

$$\left( \begin{array}{c} \displaystyle\int_0^1 \frac{2X(s)(mX(s) + b - Y(s))(1 + m^2) - 2m(mX(s) + b - Y(s))^2}{(1 + m^2)^2} \, ds \\[4mm] \displaystyle\int_0^1 \frac{2(mX(s) + b - Y(s))}{1 + m^2} \, ds \end{array} \right).$$

Hence

$$0 = \int_0^1 (X(s)(mX(s) + b - Y(s))(1 + m^2) - m(mX(s) + b - Y(s))^2) \, ds$$

and

$$0 = \int_0^1 (mX(s) + b - Y(s)) \, ds = mE(X) + b - E(Y).$$

Solving the second equation for $b$ and substituting for $b$ in the first yields

$$\int_0^1 X(s)(m(X(s) - E(X)) - (Y(s) - E(Y)))(1 + m^2) \, ds$$

$$= \int_0^1 m(m(X(s) - E(X)) - (Y(s) - E(Y)))^2 \, ds$$

and this in turn implies that

$$(m(E(X^2) - E(X)^2) - (E(XY) - E(X)E(Y)))(1 + m^2)$$

$$= m(m^2 E((X - E(X))^2) - 2m\,Cov(X, Y) + E((Y - E(Y))^2))$$

or, equivalently,

$$m\,\mathrm{Var}(X) - \mathrm{Cov}(X,Y)(1 + m^2) = -2m^2\,\mathrm{Cov}(X,Y) + m\,\mathrm{Var}(Y).$$

Hence

$$m^2\,\mathrm{Cov}(X,Y) - m(\mathrm{Var}(Y) - \mathrm{Var}(X)) - \mathrm{Cov}(X,Y) = 0$$

and using the quadratic formula, we find that

$$m = \frac{\mathrm{Var}(Y) - \mathrm{Var}(X)}{2\,\mathrm{Cov}(X,Y)} \pm \sqrt{\left(\frac{\mathrm{Var}(Y) - \mathrm{Var}(X)}{2\,\mathrm{Cov}(X,Y)}\right)^2 + 1}.$$

Since $b = E(Y) - mE(X)$, it follows that $(y - E(Y))/(x - E(X)) = m$.

**4.4.8 a)** Using the value $\overline{x}_{1100} = 192/1100$, it follows that $\alpha_{1100} \approx 0.1757$, $\beta_{1100} \approx 0.0224$ and that the corresponding 95% confidence interval is $[0.1533, 0.1981]$.

**b)** Here we find that $\varepsilon \approx 0.0224$, and therefore, the 95% confidence interval is $[192/1100 - \varepsilon, 192/1100 + \varepsilon] \approx [0.1521, 0.1970]$

**c)** $\varepsilon \le 0.98/\sqrt{1100} \approx 0.0295$, and the 95% confidence interval is $[0.1450, 0.2041]$. If we convert this result and the results in a) and b) to percentages, we find the intervals $[14.50\%, 20.41\%]$, $[15.33\%, 19.81\%]$, and $[15.21\%, 19.70\%]$. The latter two are very close to each other, but the first—the answer to c)—is a bit off. The reason for this discrepancy is that the estimate $\sqrt{x(1 - x)} \le 1/2$, which we used in deriving (4.25), is not very good if $x$ is close to zero or one, and it so happens that the value $x = \overline{x}_{1100} = 192/1100 \approx 0.1745$ is indeed relatively close to zero.

**4.4.10** Using the value $\overline{x}_{1000} = 6.13$, we find that

$$\gamma_{1000} = \frac{2 \cdot 6.13 - \dfrac{1.2816^2}{n}}{2\left(1 - \dfrac{1.2816^2}{1000}\right)} \approx 6.1393$$

and

$$\delta_{1000} = \frac{\sqrt{\dfrac{1.2816^2}{1000}\left(4 \cdot 6.13(6.13 - 1) + \dfrac{1.2816^2}{1000}\right)}}{2\left(1 - \dfrac{1.2816^2}{n}\right)} \approx 0.2276.$$

Thus, the 90% confidence interval for $\mu$ is

$$[\gamma_{1000} - \delta_{1000}, \gamma_{1000} + \delta_{1000}] \approx [5.9116, 6.3669].$$

**4.4.11** Using (4.21) with $\sigma(\mu) = \sqrt{\mu}$, we may infer that there is a 95% chance for the following inequalities to be satisfied:

$$\overline{x}_n - \frac{1.96\sqrt{\mu}}{\sqrt{n}} \le \mu \le \overline{x}_n + \frac{1.96\sqrt{\mu}}{\sqrt{n}}.$$

Equivalently, we may write

$$(\overline{x}_n - \mu)^2 \le \frac{1.96^2 \mu}{n},$$

and solving for $\mu$ yields

$$\mu \in \left[ \eta_n - \sqrt{\eta_n^2 - \overline{x}_n^2}, \eta_n + \sqrt{\eta_n^2 - \overline{x}_n^2} \right],$$

as desired.

**4.4.12** Using the value $\overline{x}_{2500} = 8.28$, it follows that

$$\eta_{2500} = 8.28 + \frac{1.96^2}{2 \cdot 2500} \approx 8.2808,$$

and therefore, the 95% confidence interval is

$$\left[ \eta_{2500} - \sqrt{\eta_{2500}^2 - \overline{x}_{2500}^2}, \eta_{2500} + \sqrt{\eta_{2500}^2 - \overline{x}_{2500}^2} \right] \approx [8.1680, 8.3936].$$

**4.5.14 b)** Using Theorem 4.5.10 in conjunction with (4.34), we need to solve for $\alpha$ the equation

$$0.015 = \int_0^\alpha f(x)\,dx$$

and for $\beta$ the equation

$$0.985 = \int_0^\beta f(x)\,dx,$$

where

$$f(x) = \frac{399 x^{397/2} e^{-x/2} 200! 2^{399/2}}{400! \sqrt{\pi}}.$$

This yields $\alpha \approx 340.1887$ and $\beta \approx 462.7550$, and since $s_{400}^2 = 4.14$, it follows that the corresponding 97% confidence interval turns out to be

$$[399 s_{400}^2 / \beta, 399 s_{400}^2 / \alpha] \approx [3.5696, 4.8557].$$

**4.5.15 b)** In solving for $y = \varepsilon \sqrt{400}/\sigma = 20\varepsilon/\sigma$ the equation

$$0.0075 = \frac{1 - 0.985}{2} = \int_{-\infty}^{-y} z(x)\,dx,$$

we find that $y \approx 2.4324$, and therefore,

$$\left[ \overline{x}_{400} - \frac{y\sigma}{\sqrt{400}}, \overline{x}_{400} + \frac{y\sigma}{\sqrt{400}} \right] \approx \left[ 18.32 - \frac{2.4324\sigma}{20}, 18.32 + \frac{2.4324\sigma}{20} \right]$$

is a 98.5% confidence interval for $\mu$. To proceed we solve for $\alpha$ the equation

$$0.985 = \int_{\alpha}^{\infty} f(x)\,dx,$$

where $f$ is defined as in the solution to Exercise 4.5.14 above. This yields $\alpha \approx 340.1887$ (as in Exercise 4.5.14), and therefore,

$$[0, 399s_{400}^2/\alpha] \approx [0, 4.8557]$$

is a 98.5% confidence interval for $\sigma^2$. Consequently, an upper estimate for a 97% confidence interval for $\mu$ is

$$\left[18.32 - \frac{2.4324\sqrt{4.8557}}{20}, 18.32 + \frac{2.4324\sqrt{4.8557}}{20}\right] \approx [18.052, 18.588].$$

**4.6.6** According to Theorem 4.6.3, we need to solve for $\varepsilon$ the equation

$$0.99 = \int_{-\varepsilon}^{\varepsilon} f(x)\,dx,$$

where

$$f(x) = \frac{\Gamma(450)(1 + x^2/899)^{-450}}{\Gamma(899/2)\sqrt{899\pi}}.$$

This yields $\varepsilon \approx 2.5813$, and therefore, a 99% confidence interval for $\mu$ is

$$\left[13.15 - \frac{\varepsilon\sqrt{2.46}}{30}, 13.15 + \frac{\varepsilon\sqrt{2.46}}{30}\right] \approx [13.015, 13.285].$$

# Bibliography

[A]     Ash, Robert B., *Real Analysis and Probability*, Acadmic Press, San Diego, 1972.

[B1]    Blume, Frank, *Applied Calculus for Scientists and Engineers*, Volume 1, Createspace, 2014.

[B2]    Blume, Frank, *Applied Calculus for Scientists and Engineers*, Volume 2, Createspace, 2014.

[B3]    Blume, Frank, *Science and Spirit*, Createspace, 2014.

[D1]    Dunne, Brenda, Gender Differences in Human/Machine Anomalies, *Journal of Scientific Exploration*, Vol. 12, No. 1, pp.3–55, 1998.

[D2]    Dunne, Brenda, Co-Operator Experiments with an REG Device, *PEAR Technical Note 91005*, December 1991.

[JD1]   Jahn, R.G., Dunne, B.J., *Consciousness and the Source of Reality*, ICRL Press, Princeton, New Jersey, 2011.

[JD2]   Jahn, R.G., Dunne, B.J., *Margins of Reality*, Harvest Book, San Diego, 1987.

[JDN]   Jahn, R.G., Dunne, B.J., Nelson, R.D., Dobyns, Y.H., Bradish, G.J., Correlations of Random Binary Sequences with Pre-Stated Operator Intention: A Review of a 12-Year Program, *Journal of Scientific Exploration*, Vol. 11, No. 3, pp.345–67, 1997.

[K]     Körner, T. W., *Fourier Analysis*, Cambridge University Press, Cambridge, 1995.

[McT]   McTaggart, Lynne, *The Field*, HarperCollins Publishers, New York, 2002.

[S1]    Schmidt, Helmut, Mental Influence on Random Events, *New Scientist and Science Journal*, June 24, 1971, pp.757–8.

[S2]    Schmidt, Helmut, Quantum Processes Predicted?, *New Scientist*, October 16, 1969, pp.114–5.

# Index